ALSO BY CHARLES A. WHITNEY

The Discovery of Our Galaxy (1971)

THIS IS A BORZOI BOOK, PUBLISHED IN NEW YORK
BY ALFRED A. KNOPF

Whitney's Star Finder

WHITNEY'S

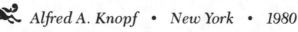

Alfred A. Knopf • New York • 1980

STAR FINDER

A Field Guide to the Heavens
Revised Edition

CHARLES A. WHITNEY

THIS IS A BORZOI BOOK
PUBLISHED BY ALFRED A. KNOPF, INC.

Grateful acknowledgment is made to Dover Publications, Inc.,
for permission to reprint material from Canon of Eclipses
(Oppolzer); W. H. Freeman and Company for a letter from
Adm. G. L. Schuyler (pages 148–50) in the "Amateur Scientist,"
March, 1956. Copyright © 1956 by Scientific American, Inc.
All Rights Reserved; and to the Lick Observatory, University of California,
Santa Cruz, for use of a photograph of the Great Nebula of Orion.

Library of Congress Cataloging in Publication Data
Whitney, Charles Allen.   Whitney's Star finder.
Includes index
1. Astronomy—Observers' manuals.   I. Title.   II. Title: Star finder.
[QB65.W48 1977]   523   77–75359
ISBN 0–394–73405–X

Manufactured in the United States of America

Published October 3, 1977
Reprinted Three Times
Fifth Printing, August 1980

*Dedicated to the memory of
my brother,
who taught me how to skin a snake*

Contents

1: WHAT IS THAT BRIGHT OBJECT?   3

Streets and traffic in the sky; Finding north; Using the star finder to identify stars; Can it be a planet?; Colors and brightnesses of the planets; Airplanes and artificial satellites

your eyes and binoculars on double stars; Variable stars; The Orion nebula; The zodiacal light; Star clusters; The Andromeda galaxy (M31)

APPENDIXES

List of Figures

List of Tables

Preface

This is a field guide to the sky. It does not attempt to explain astronomical knowledge or the ways in which astronomers study the sky. It attempts to explain the sky itself.

Friends often ask me, "What was that bright thing I saw last night?" I have written this book for them, and I hope it will help them find the answer right then, when they can take another look—and before they need to put me on the spot. I have limited myself to objects that are visible to most campers or might be of interest to those who merely wish to step outside for a few minutes with a small telescope or a pair of binoculars.

A special feature is the star finder in the back of this book. With this you will be able to predict the positions of the stars for any hour of any day, and you will be able to tell time by the stars and predict the time of sunset.

This book will also help you see the moon and Venus in broad daylight or catch a glimpse of the Andromeda galaxy and the Orion nebula. If you wish to know why Venus becomes bright in the evening sky and then jumps to the morning sky, turn to Chapter 3. That chapter will also help you see the moons of Jupiter with binoculars. The appendixes provide lists of coming events in the sky.

To cast your horoscope, locate the planets with the information in the Almanac of the Planets, and ask someone what it all means. This book will not help you interpret your horoscope. Sorry.

C. A. WHITNEY

Weston, Massachusetts
September, 1973

Preface to the First Revised Edition

I am very grateful for the enthusiastic reception to the first edition, and I have revised the material in three ways. First, I have updated the tables and descriptions of events and planetary appearances. Second, in response to comments from readers, I have added descriptions of several phenomena that were not included in the first edition. Third, I have outlined some new information gleaned from interplanetary flights. These flights have made us feel more at home among the stars, as I hope this book will continue to do for its readers.

C. A. WHITNEY

Weston, Massachusetts
March, 1977

How Your Location Affects the Use of This Book

In a word, your location affects the use of this book very little, and there are many aspects of the sky which would be unaffected by your taking a trip around the world.

Suppose you started from Boston at noon and flew eastward at 750 miles per hour. This velocity would take you around the circle of latitude and return you to Boston in twenty-four hours, but you would have seen two sunrises during that time. Your day would seem half as long as an ordinary day for a person on the ground. On such a trip the North Star would stay at the same height in the sky, and the familiar constellations would be visible. In other words, eastward travel merely takes you into other time zones.

And if you traveled *westward* from Boston at 750 miles per hour, the sun would remain overhead the entire time. Over the Pacific Ocean you would cross the International Date Line and would have to tear a page off your calendar. In twenty-four hours you would have returned to Boston and a new day would have arrived, but you would have seen no sunset.

Suppose, on the other hand, you went to the North Pole. The North Star would be directly overhead, and the constellations would circle the horizon without rising or setting. The sun would rise and set once a year, and the moon once a month. Half of the sky would be perpetually invisible, and you would never see Orion, for example, because it lies below the equator.

Now imagine you head for the South Pole. The North Star gradually sinks toward the northern horizon—one degree for every 67 miles—and new constellations appear. As you cross the equator, the North Star vanishes and you notice that the stars are rotating about the South Pole of the sky, but clockwise rather than counterclockwise as they do in the north. Also, you will find winter when you had previously found summer, and spring in place of autumn.

Although the star background changes as you travel south, several aspects of the sky remain unaltered. First, the sun would continue to cross your meridian at noon (or very nearly at noon) on your watch every day. You would remain in the same time zone. Also, the phases of the moon would be unaltered by your travel, and eclipses would occur on the same nights. The planets would remain among the same stars, and Venus, for example, would still be a morning or evening "star," as it had been in the other hemisphere. You might notice that the moon had moved a degree or two northward, but none of the planets would be sensibly shifted unless you observed them with a powerful telescope.

Thus, although I live near Boston and have written this book for the "hub of the universe," much of it applies to other places, and when a correction is necessary the procedure is fairly simple—if you follow me. If you don't, you can often ignore the corrections without much danger. Good luck.

Whitney's Star Finder

What Is That Bright Object?

STREETS AND TRAFFIC IN THE SKY

Learning one's way among the stars is like learning one's way through the streets of an unfamiliar city. We start at a prominent intersection and move along boulevards and thoroughfares, progressing gradually to the lesser streets and alleys. We look for landmarks, but we ignore cars, trucks, and pedestrians, because many of them will have moved before we return to the same spot.

In much the same way, many constellations of stars are easily recognized: Orion, the Big Dipper, the W of Cassiopeia, the Pleiades cluster. These may serve as landmarks, but the moon and the planets are pedestrians, and we often find them in unexpected quarters. Comets are the foreigners, and they appear in strange clothing. Their behavior seems eccentric, and although they are inclined to avoid other pedestrians, they respond enthusiastically to the light of the sun. Meteors are their gypsy children, brilliant and unpredictable. They move in packs and rarely leave the orbits of their parents, nor do they often touch ground.

This chapter discusses the first step in reading the map of the sky: orienting yourself so you will not be caught holding the map upside down.

FINDING NORTH

Because the North Star (Polaris) is nearly stationary in the sky, the ancient Egyptians thought it pointed the way to the realm of perpetual life. For the same reason, the North Star is the best reference point for orienting yourself on the sky. It is fairly bright, and it stands alone about halfway to the zenith from the northern horizon, forming the tip of the handle of the Little Dipper. If you find this constellation, you can be sure of your identification, but it is not a very bright group of stars, and I usually find the North Star in other ways.

The easiest way is to use the Big Dipper. If you find it, your problem is solved, because the end stars in the bowl of the Big Dipper always point directly to the North Star, as you can verify with the star finder in this book.

There are many other clues, and these may be useful if the Big Dipper is hidden. In particular, there are many constellations that have a well-defined "north side." The top of Orion, for example, always indicates the direction of the North Star. And if you find Cassiopeia and turn your head so it looks like a proper W, you will find the North Star toward the top of your head. Leo, the Lion, lies in the sky so that the North Star is above his back. You can discover a number of other indicators.

Even without using any of the constellations, you can find north if you lie on your back under a tree for a few minutes and watch a star near the zenith.

(The tree branches make a good reference.) Within a few minutes, you will be able to detect the motion of this star caused by the rotation of the Earth. The star moves westward.

Orientation is quite simple on a sunny day. At noon, standard time, the sun is in the south, but if you would rather not wait until noon you can find south in a few minutes with a method borrowed from the *Scout Handbook*. This method uses the westward *motion* of the sun rather than the *position* of the sun at a particular time, and it will work well enough at any time of day, although it is most accurate around noon.

Find a pointed stick about four feet long. If you are in a rush, use one eight feet long—the shadow will move twice as fast. Jam it into the ground so that it points more or less upward and put a small stone at the tip of the stick's shadow. Now wait a few minutes. As the sun moves west, the shadow will move directly east, and when it has gone a few inches, put another stone at the tip of the shadow. The two stones lie east and west of each other. (That's not surprising. What *is* surprising is the speed of the shadow. If you did this using the shadow of the Eiffel Tower, the shadow would move seven feet per minute!)

USING THE STAR FINDER TO IDENTIFY STARS

This book contains a star finder which may be set for any hour of any day of the year. If you know the time and have found north, you are in a position to use the star finder to identify the stars. Set the date against the appropriate hour and hold the star finder over your head with the edge marked "south" pointing toward the south. In this position, it provides a map of the sky, and you will be able to pick out the constellations and the bright stars.

The stars return to the same positions each year. This is not true of the planets, and because they move among the stars, they are not shown on the finder. They must be handled in a different manner.

CAN IT BE A PLANET?

If a bright object is stationary and does not appear on the star finder, it may be a planet. With a telescope, you can easily identify the planets because each shows a disk, and the disks are all different. (The appearance of each planet is discussed in Chapter 3.) Without a telescope, it is often easier to tell what is *not* a planet than what *is*. Here are several ways you may be able to eliminate the planets from your list of suspects.

Does it twinkle? If the object shows strong twinkling, it probably is not a planet. Except for Mercury, the planets do not twinkle. (But neither do the stars when they are high overhead.)

Is it near the North Star? If so, it cannot be a planet. This is probably the one rule in this book that needs no qualification and has no exceptions. (Planetary orbits follow the path of the sun and never come within 50° of the North Star.)

Is it high in the south during summer? If so, and if you are observing from the northern half of the United States, the object is probably not a planet.

Is it very low in the south during winter? The answer to the previous question applies here also. Planetary orbits take them fairly high in the sky during winter. Of course, this is not true when they are rising and setting, so watch your directions because planets rise in the east or southeast and set in the west or southwest.

COLORS AND BRIGHTNESSES OF THE PLANETS

Each planet has a distinctive color, and although the tints are not particularly striking, they are useful for identification. The colors are as follows:

PLANET	COLOR
Mercury	Leaden
Venus	Silvery
Mars	Reddish
Jupiter	White
Saturn	Yellowish

If you wish to make definite identification of a planet, first determine the constellation in which you see it, and then turn to the Almanac of the Planets in the back of this book. When two planets are in the same constellation you can distinguish them by their colors.

In addition to having a characteristic color, each planet also has a characteristic brightness. Although these brightnesses vary somewhat from season to season, they are helpful in identifying the planets. The following list compares each planet to Vega and Rigel, bright stars of the summer and winter sky, respectively.

PLANET	BRIGHTNESS RELATIVE TO VEGA AND RIGEL
Mercury	Similar
Venus	40×
Mars	4×
Jupiter	6×
Saturn	Similar

AIRPLANES AND ARTIFICIAL SATELLITES

Except for their motion, these objects often appear starlike. Binoculars will usually reveal the red and green wing lights of an airplane, but otherwise these objects are not easily distinguished. Both can show variations of brightness, and both can cross the sky in fifteen minutes or so. There are two ways to tell them apart. Airplanes can turn corners, while satellites rarely do. Also, satellites can vanish while high overhead, but airplanes rarely do. On a summer night, or near twilight in winter, you may see a satellite vanish into the shadow of the Earth. This is an eclipse, of sorts, and it is abrupt.

Finding the Moon

THE MONTH

Suppose the Earth stopped rotating and the sun became fixed in the sky on the western horizon just after sunset. The moon would then rise in the west once a month, move upward, and then set in the east. Because the full circle requires 29½ days, the moon would move about 12° eastward per day, slightly less than the spread of your fingers when you stretch your arm full length. At the time of the new moon, it passes within 5° of the sun—or less, depending on the orientation of the moon's orbit that month. About one day after new moon, the thin crescent appears in the evening twilight.

SEEING THE NEW MOON

We seldom notice the new moon until it is at least two days old, but with a little luck and a clear sky it can be seen a day earlier. Twenty-one hours after the time it passes nearest to the sun is the earliest you can expect to see it, and even then you will see it only for a short period after sunset. Binoculars will help.

SEEING THE MOON ANY DAY

Except for a day or two on either side of new moon, you can find the moon any day of the month by consulting the Phases of the Moon table in the back of this book.

When the new moon is "old" enough to be visible (a day or two), it will be found slightly east of the sun. It then sets shortly after sunset. One week after new moon, the moon is at "first quarter." Its face is half-illuminated and it is 90° east of the sun, so it rises in the east at noon and it sets at midnight. You will find it in the southeast portion of the sky during the afternoon.

On the night of full moon, you will see it opposite the sun; it rises at sunset and sets in the west at sunrise. Thus, you will not see the full moon in the daytime sky.

The full moon gives only 1/500,000 the light of the noonday sun, so a camera capable of photographing scenery with an exposure of 1/1000 second during the day will require 500 seconds, or nearly eight minutes, in bright moonlight.

Two nights after full moon, the light is only half as bright, and this rapid drop is caused by the peculiar reflecting properties of the moon's surface. Like a highway sign covered by reflecting buttons, it throws light back toward the sun, so it appears much brighter when the sunlight comes directly over your shoulder on the night of full moon.

One week after full moon, the moon is at third quarter, and it will be found 90° west of the sun. It then rises at midnight, is due south at sunrise, and sets at noon. You will find the third-quarter moon in the southwest portion of the sky during the morning.

With the table of lunar phases at the back of the book, you may estimate the phase of the moon on any day of the year. First quarter is one week after new moon, and third quarter is one week after full moon.

The phases of the moon do not repeat exactly from one *calendar* month to the next because the average length of the calendar month is one-twelfth of a year, or 30½ days, while the phases repeat every 29½ days. Thus, the full moon comes about one day earlier each month, on the average.

MOONRISE

The moon seldom follows precisely its "average" behavior. If it did, it would rise 50 minutes later each day, but the interval actually varies from 30 minutes to 75 minutes, and this variation makes it very difficult to guess the time of moonrise. Rather than give a complete table, I shall merely indicate some general rules, because newspapers often provide the times of moonrise and moonset with the weather forecast.

The *position* of the moon's rising will influence the *time* of its rising. The moon moves north and south across the celestial equator during its monthly motion eastward around the Earth. The total swing north and south is 45°, and the moon seldom rises in the exact east. When the moon is moving southward, its time of rising is delayed 75 minutes each day, and when it is moving northward the delay is only 30 minutes. (This rule is reversed in the Southern Hemisphere.)

In late winter and early spring, the full moon is moving rapidly southward, so the time of rising is delayed by its greatest possible amount. In late summer and early autumn, the full moon is moving rapidly northward, so its delay in rising is near the smallest possible value.

One result of this behavior is the "harvest moon," which is the full moon nearest the autumnal equinox. For a few days, the moonrise is delayed only 30 minutes per day, and the "hunter's moon" one month later shows a similar pattern. By contrast, the autumnal new moon's rising is delayed by the maximum amount each day, because it is similar to the spring full moon.

HEIGHT OF THE FULL MOON

In late spring and early summer, the full moon stands low in the southern sky at midnight. It rises much higher during the late autumn and early winter.

The cause of this behavior is the tilt of the Earth's axis and the fact that the full moon lies opposite the sun in the sky. When the sun is farthest north (June), the full moon is farthest south. And when the sun is farthest south (December), the full moon is farthest north.

There is a slight variation on this rule, caused by the 5° tilt of the moon's orbit with respect to the path of the sun. In some years, this tilt enhances the north and south motion of the moon; in other years it reduces it. The cycle is eighteen years, but the amount (10°) is hardly noticeable.

ASHEN LIGHT OF THE MOON

Near the time of new moon (either morning or evening), the full face of the moon can usually be seen within the thin crescent. This is called the "old moon in the new moon's arms," or the "ashen light." To understand it, imagine you are standing on the surface of the moon at this time. You will see a "full Earth," that is, the face of the Earth will be fully lit by the sun, and it will be more than ten times as bright as our full moon. The "moonscape" around you will be so brightly lit that it will be visible from the Earth as the "old moon in the new moon's arms." Careful observation of this ashen light shows that it varies from time to time. It is brightest on the evenings when the moon side of the Earth is most covered with clouds, which brightly reflect sunlight into space.

Observing the Planets with Binoculars

THE APPEARANCE AND MOTION OF THE PLANETS

The planets shine by light reflected from the sun, and one or more of them can be seen on most clear nights. Their apparent positions change very little from one part of the Earth to another, but over a period of a week it is possible to detect their motion if you locate them carefully among the stars.

The solar system is remarkably flat, and the planets remain within a narrow path, the "zodiac," because they lie within a single plane, like marbles resting on the top of a table—or nearly so. This plane also contains the Earth and the sun, and if we viewed it edge-on from one of the stars of the zodiac, we would see the planets strung out in a line. Only if we were to see the planets from a star far above or below the zodiacal circle would we see the roundness of their orbits about the sun.

The flatness of the solar system and the roundness of the orbits suggest that the planets were formed from a nebula rotating about the sun. According to the most popular theory, solid particles within the nebula came together to form larger particles and sweep up the drifting gases. Only the largest survived. Around each of them a similar swarm of particles formed the satellites. (The theory that the planets were formed from gas torn out of the sun by a passing star is now considered by most astronomers to have too many difficulties, principally because such matter would not condense into planets, and even if it could, the planets would fall back into the sun.) If current thinking is correct, the formation of planets around stars may be a fairly common occurrence.

The Earth is the third of the known planets: Mercury, Venus, Earth, Mars, Jupiter, Saturn, Uranus, Neptune, Pluto. The first six were known to ancients, and the remaining three were discovered in modern times by a combination of computation, acuity, and luck. William Herschel discovered Uranus in 1781 while scanning the sky with a telescope. He thought he had discovered a comet until its subsequent motion proved otherwise. The existence of Neptune was postulated to explain the peculiar irregularities in the motion of Uranus, and it was discovered in 1846. In a similar fashion, irregularities in the motion of Neptune stimulated a search for a ninth planet, and the search ended with Clyde Tombaugh's discovery of Pluto in 1930. (Oddly enough, the predictions on which Tombaugh based his search were later shown to have been incorrectly computed, although they gave the correct result.)

Peculiarities in the motion of Halley's Comet were recently interpreted as indicating the presence of a tenth planet, beyond Pluto, but further analysis showed that the hypothetical planet would have disturbed the other planets out of their present orbits. The conclusion is that Halley's Comet is disturbed by other forces, perhaps by the impulse of gas being ejected from its body.

The last three planets are invisible to the ordinary observer, so I will ignore them for the remainder of this book. (Actually, Uranus is slightly brighter than the limit for the naked eye, but I have never succeeded in seeing it—partly because I haven't tried very hard.)

Planets nearer than the Earth to the sun behave quite differently from the outer planets. The inner planets remain fairly close to the sun because we see their orbits from the outside. The maximum angles are 28° for Mercury and 46° for Venus. The outer planets, on the other hand, may be found at any angle from the sun.

Another feature distinguishing the behavior of these two groups is that the inner planets can show crescents like the moon's, while the outer planets always appear nearly full.

All planets move about their orbits in the same direction: counterclockwise when seen from the North Star. This is also the direction of the Earth's daily rotation and the rotation of the other visible planets with the exception of Venus.

The outer planets move much more slowly than the inner planets. (Their years are longer, and they move fewer miles per hour.) There is a regular progression, explained by Isaac Newton in terms of the sun's gravitational field, so if the planets were one day lined up like spots on the spoke of a wheel, the inner ones would quickly move ahead. Thus, the planets are continually rearranging themselves.

JUPITER AND ITS MOONS IMITATE THE SOLAR SYSTEM

Jupiter has thirteen moons (the last was discovered in 1974, and it has a diameter of only two miles). Four of them would be visible with the naked eye if it were not for the glare of Jupiter's light, and they can often be seen with a good pair of binoculars mounted on a very steady support.

The moons appear to lie on a straight line and to move back and forth along this line. Thus, the number of moons on each side of Jupiter will vary from night to night. The two inner moons move rapidly, and their motions can be seen during a single night, but the outer two move very little from one night to the next. The period of each moon is determined by the mass of Jupiter and by the moon's distance from Jupiter. (The moon's mass is irrelevant as long as it is small.) A similar relationship holds for the planets; their orbital periods depend only on the mass of the sun and the planet's distance from the sun. These relationships are explained by Newton's law of gravity.

MERCURY

Relatively few people notice Mercury. Its orbit size is only two-fifths that of the Earth's orbit, and Mercury is never more than 28° or about two hand-widths from the sun. It must be hunted in the twilight.

Look for Mercury within a week of its greatest eastward (evening) or westward (morning) elongations from the sun. It is then seen as a zero-magnitude star, similar to Vega, and it can be spotted quite easily with a pair of binoculars. You will not see its disk without a magnification of at least 50×.

Mercury is more easily seen at some elongations than at others. The most favorable times are morning elongations in the fall and evening elongations in the spring. (The dates of elongations are listed in the back of this book in the Almanac of the Planets.) Two factors affect the visibility of this planet from season to season. First, its orbit is highly elliptical and its true distance from the sun varies by 50 percent. Second, and more importantly, the planetary path around the sky is more steeply inclined to the horizon during autumnal sunrises and spring sunsets. See Figure 1.

Mercury's markings are very difficult to see from Earth, even with a large telescope, but the *Mariner 10* fly-by revealed numerous meteor craters, very much like those on our moon. Its magnetic field is about one-hundredth that of the Earth, and it has a very thin atmosphere which it may have captured from gases flowing out from the sun. Mercury's daytime temperature is hot enough to melt lead, and even at night it is warm enough to boil water.

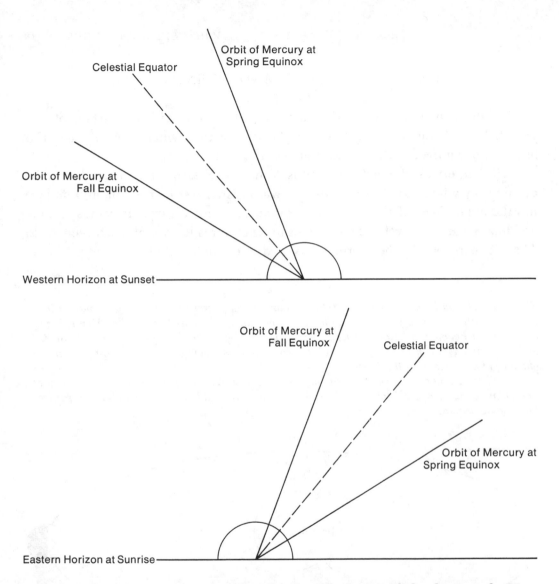

Orbit of Mercury at
Spring Equinox

Celestial Equator

Orbit of Mercury at
Fall Equinox

Western Horizon at Sunset

Orbit of Mercury at
Fall Equinox

Celestial Equator

Orbit of Mercury at
Spring Equinox

Eastern Horizon at Sunrise

Figure 1. Schematic representation of the fall and spring sky at twilight, showing why Mercury is most easily seen after sunset in the spring and before sunrise in the fall. Mercury's height above the horizon depends on the angle between its orbit and the horizon, as well as the planet's distance from the sun. As indicated here, the orbit makes a greater angle with the horizon during spring sunsets and fall sunrises. These are usually the best times to glimpse Mercury.

VENUS, THE EVENING AND MORNING STAR

Venus comes nearest to the Earth, and it has the brightest surface of any planet. Its brilliant silver light is unmistakable in the twilight, and it may also be seen during the day throughout much of the year.

At intervals of nineteen months Venus reaches maximum height in the evening sky when it is at its greatest eastward distance from the sun, and its face has the appearance of the moon at first quarter. A month later it begins to fade and finally vanishes in the glare of the sun, where it is lost for about three weeks. Then it reappears in the morning sky and rises higher each day before being

Figure 2. Schematic view from the Earth showing the appearance of Venus as it moves in its orbit about the sun. The portion illuminated by the sun is crescent-shaped when Venus is near the Earth; it becomes a half-circle when Venus is at its greatest elongation from the sun; it is gibbous when Venus is on the far side of its orbit. Note the changes of size accompanying the changes of distance from the Earth.

These changes of appearance, which may be seen with a securely mounted pair of 7× binoculars or with a small telescope, formed one part of Galileo's argument for the sun-centered solar system.

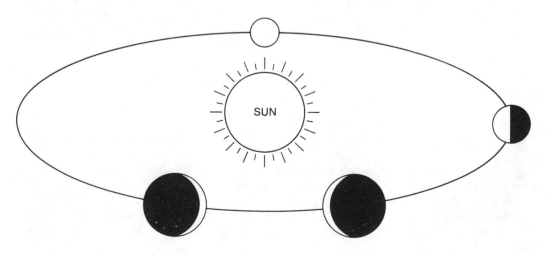

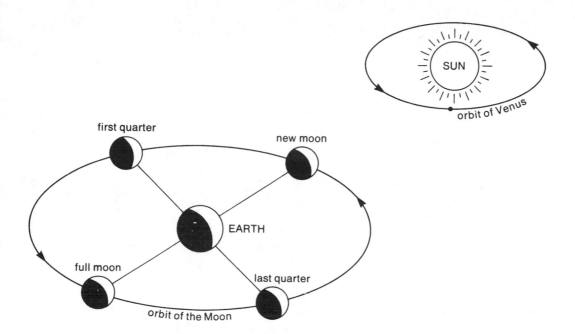

Figure 3. Schematic representation of the moon's orbit around the Earth and Venus's orbit around the sun. Phases of the moon are indicated. Because the moon's orbit circles the Earth rather than the sun, the moon always appears as a crescent when it is near the sun. Compare with the behavior of Venus indicated in the preceding figure.

lost in the glare of the rising sun. Two months later, it is at its greatest westward distance from the sun and it stands high in the morning sky at dawn. Then it again vanishes into the glare of sunrise and remains invisible for two months, reappearing in the evening sky and repeating the cycle.

Venus shows a changing crescent that can be seen quite well with a telescope magnification of 7× or 10×. Binoculars will work if they are securely mounted. This crescent has the same origin as that of our moon, but the two crescents will generally have different shapes, even when they are seen in the same part of the sky. For example, if you see Venus in the evening sky near the thin crescent of the new moon, Venus may have either a full face or a thin crescent.

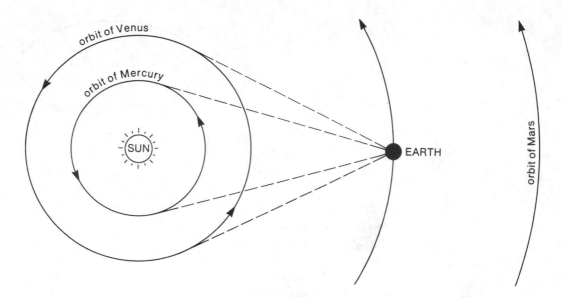

Figure 4. The orbits of Mercury, Venus, Earth, and Mars, seen from above the solar system. Because Mercury and Venus are closer to the sun than the Earth is, they remain near the sun in the sky. Mars, in an outer orbit, occasionally lies opposite the sun—approximately every two years.

The reason for this difference is that Venus may lie on the opposite side of the sun, while the moon may not. See Figures 2 and 3.

The silvery luster of Venus is produced by a thick, hot atmosphere of carbon dioxide nearly one hundred times as dense as the atmosphere of the Earth. Winds of 200 miles per hour blow continually around the planet, so rapid that there is no difference between daytime and nighttime temperature near the surface. Radar signals have revealed extensive mountain ranges beneath the clouds.

INTERPRETING THE MOTIONS OF VENUS AND MERCURY ON THE SKY

Suppose the daily rotation of the Earth were stopped and the sun stood still in the sky. Venus and Mercury would move back and forth along the zodiac, nearly crossing the face of the sun as they move from east to west and passing behind the sun as they move from west to east. (On rare occasions, these planets do cross the sun's face and are seen as tiny black disks.)

If we were far above the solar system looking down on the North Pole of the Earth, the orbit of Venus would resemble the circle in Figure 4. The orbital motion of Venus and the Earth, as well as the daily rotation of the Earth, are counterclockwise in this figure, so Venus will be seen in the evening sky when it is on the left hand side of the sun. (Think about that for a minute. Eastward is counterclockwise in this view, so evening objects are to the left of the sun.) When Venus overtakes the Earth and passes nearly in front of the sun (inferior conjunction), it vanishes for about three weeks in the twilight and reappears in the morning sky—to the west of the sun. Thus, the moment of its passage from evening to morning sky is also the moment of its closest approach to the Earth. And it is the time of its thinnest crescent.

SEEING VENUS AND SIRIUS IN THE DAYTIME

Unlike any other planet, Venus may be seen by the naked eye in full daylight for several months around the times of its greatest apparent distance from the sun.

For my first attempt at seeing Venus in the daylight, I watched it in the morning twilight just before dawn and measured its distance from the sun using my extended fingers. (This distance remains constant as the two objects move across the sky during the day.) Later in the morning, I swept the suspected location of Venus with a pair of binoculars, and when I found it I pinpointed its location against the branches of a tree. Then I lowered my binoculars and, in a moment, the planet jumped out at me. I wondered why I had not seen it sooner.

Sirius, the brightest star in the sky, is about one-tenth the typical brightness of Venus, and it may also be seen in full daylight, but you'll need a telescope. If you point a three-inch, or greater, telescope accurately at the location of Sirius, you will see it without much difficulty. I have tried to find Sirius with a 1½-inch pair of binoculars, but without success.

MARS

Mars's unblinking reddish light makes it fairly easy to find among the stars. The orbit of Mars lies outside the Earth's and it can be seen directly opposite the sun every two years, when it is closest to the Earth. Even then, its disk is only one-hundredth that of the moon, so a good pair of binoculars will barely show the disk. A telescope with a magnification of 200✕ is required to see its surface markings.

A close approach of Mars to the Earth ("opposition") will occur in January, 1978.

Mars's motion on the sky is quite different from that of Venus and Mercury. At opposition it rises at sunset and can be seen all night. Two months before opposition it rises at midnight and can be seen in the southeastern part of the sky for several hours before dawn. Two months after opposition it is seen in the evening sky and it sets at midnight. (Jupiter and Saturn do much the same, but their oppositions occur nearly every year.)

The motion of Mars among the bright stars can be seen by the naked eye if you watch for a week or so. This motion appears rather erratic.

For most of the year, Mars moves eastward among the stars at a rate of about 20° per month, but when it approaches opposition it slides backward about 15° in two months. This is its "retrograde motion," and it is caused by the orbital motion of the Earth. It is a consequence of our perspective. If the Earth were stationary (an impossibility in the neighborhood of the sun), Mars would move continually eastward and circle the sky in twenty-three months. Because the Earth moves fairly rapidly, Mars is overtaken every twenty-six months, and it falls behind for a while.

MODERN VERSION OF THE "CANALS" OF MARS

Photographs of Mars from space satellites and landers have revealed enormous volcanic craters, and many long valleys resembling dried river beds. Water erosion was very extensive at some time in the past, and all of the water now appears to be trapped in the polar caps, and perhaps under the ground. None of these features explains the "canals" reported at the beginning of this century. They were evidently the result of optical illusions. The surface of Mars is reddish and strewn with broken rocks. Even the sky appears pink from the ground.

Mars is 100° F cooler than the Earth because it is farther from the sun, and this difference will some day make Mars an attractive place to live, because the sun is growing brighter and will warm the Earth beyond the comfort point in a few billion years—according to current theories.

JUPITER

Jupiter excels all other planets in size, number of moons, speed of daily rotation, markings, and availability to the observer with binoculars or a small telescope. (The ring of Saturn is more bizarre, but that is Saturn's only claim to superiority.) Jupiter is second to Venus in brightness. Its orbit about the sun is five times larger than the Earth's and requires twelve years, so the planet moves along at an average rate of one zodiacal constellation per year. Thus, Jupiter will be seen at opposition about one month later each year. Like Saturn it may be seen in the evening sky five months of the year and in the morning sky five months. The remaining two months are lost to the light of the sun.

Jupiter rotates so rapidly (once in ten hours) that a patient observer with a 5-inch or 10-inch telescope can see the movement of its clouds in a few hours. It has an intense magnetic field and its upper atmosphere emits strong radio waves. One of the most amazing discoveries by the *Pioneer* space probes is that

Jupiter sends out two or three times as much heat as it receives from the sun. The sources of this heat, and of the Great Red Spot on the clouds, are still unknown.

SATURN

Saturn moves more slowly than Jupiter, and it circles the zodiac in thirty years, so you will see it in the same constellation for more than a year at a time. Also, its orbit is nearly circular and ten times larger than the Earth's, so its distance from us changes relatively little. Saturn shows no crescent (nor do Mars and Jupiter, for that matter), and it always appears nearly the same size in a telescope.

The moons of Saturn are invisible with binoculars, and its surface is hidden beneath a layer of clouds. Galileo's telescope was not much better than present-day binoculars, and when he first examined Saturn he thought it had "ears" or lobes on each side. Part of Galileo's difficulty was that he did not *expect* to see rings. You will probably have a somewhat easier time of it, but magnification of 25× or more is required to see the rings clearly.

The spectroscope shows that the inner portion of the ring moves more rapidly than the outer, so it cannot be a solid. Its reflecting properties suggest that it may be composed of snowflakes and bits of dust. Although the ring is twice the diameter of Saturn, it vanishes when viewed edge-on, so it must be very thin. Several astronomers have claimed the ring is only a few inches thick! (And some astronomers have believed them.)

Determining Time
by the Sun and Stars

NATURAL CLOCKS

The sky provides a number of clocks, and the most obvious is the sun. Our *civil* clocks (the kind we find in grocery stores) are set to sun time, and their hour hands rotate twice while the sun rotates once around the sky. The rotation of the moon about the Earth defines the month, and the rotation of the Earth about the sun defines the year. (Longer cycles can also be discerned. For example, the rotation of the sun about the center of the Milky Way defines the "cosmic year" of about 200,000,000 terrestrial years.)

This chapter will consider three types of time determination: the hour during the day, the hour during the night, and the season of the year. They depend only on the motion of the sun and stars, and there will be no further mention of the planets or the moon in this chapter.

SUN TIME AND TIME ZONES

Civil clocks mark twenty-four hours while the sun makes a complete revolution of the sky (360°), so the sun moves 15° per hour. (See Figure 5.) For

Figure 5. "Solar time" is determined by the position of the sun in the sky. In twenty-four hours, the sun moves 360°, or 15° each hour.

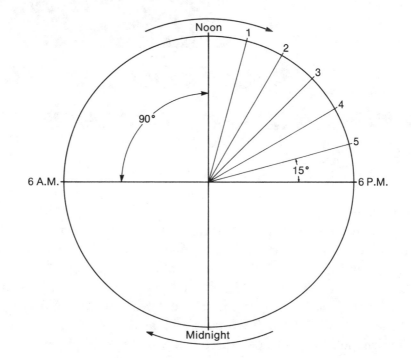

an observer at the center of his time zone, the sun is directly south at noon; it is 90° eastward at 6 A.M.; it is 90° westward in its cycle of the sky at 6 P.M. (This assumes standard time. During daylight-saving time these events occur one hour later.[1])

For an observer who is not at the center of his time zone, the situation is slightly different. If he is *east* of the center, the events occur at an *earlier* civil time. For example, Cleveland and Boston are in the Eastern Time Zone, but the sun rises forty minutes earlier in Boston, because Boston is 10° of longitude eastward, toward the rising sun.

The time zones are spaced at 15° intervals around the globe, and the sun moves from one to the next in an hour. Boston is 71° west of Greenwich, England (where it all begins), so it is in the fifth zone. But it is 4° east of the center of its zone (75°), so its sun time is four-fifteenths of an hour, or sixteen minutes, ahead of Eastern Standard Time. Therefore the sun comes due south sixteen minutes earlier than our civil clocks would indicate. Knowing your own longitude, you can estimate the correction from civil time to sun time at your location, remembering *eastward* means the sun is early.

MAKING A SUNDIAL

The mathematical design of an accurate sundial is a fascinating and complex problem, but the practical procedure is very simple. It requires nothing but time, and the accuracy you achieve is mostly a matter of size. If you make yours large enough, you can time the boiling of eggs.

For the moment, I will assume you have a watch, because it simplifies the procedure, but I will later mention how it could be done with a compass.

---

[1] There is a simple way to recollect the difference between standard and daylight-saving time: The clock *springs* forward to daylight-saving time in the spring, and it *falls* back to standard time in the fall.

(Actually, you don't even need a compass, but again it simplifies matters if you have one.)

Find a thin stick about six feet long. Jam it into a piece of flat ground where the sun will shine most of the day, and collect a bunch of small sticks. Carve an hour number on each, and when your watch indicates that an hour has arrived, put the appropriate stick at the tip of the shadow of the large stick. Do something else for an hour, then put in another stick, and do this all day—or do it every half-hour or quarter-hour if you want more accuracy. If your stick is six feet long, the shadow will move about two feet east every hour, and more rapidly near sunset and sunrise.

This sundial will work all year, but as the seasons pass, and the sun's height increases and decreases, the shadow will fall at varying distances from the foot of the stick. You can put more hour sticks in after a week or so, and if you mark them with the date you will have a calendar as well as a clock.

A SUN CALENDAR-CLOCK IN YOUR HOUSE

An indoor sundial was described in the March, 1956, issue of *Scientific American* by Rear Admiral Garret L. Schuyler, U.S.N. (Ret.), of Washington, D.C. It takes advantage of a long sunbeam and is therefore more precise than most sundials, and it can be constructed on the ceiling of a bedroom—a clear advantage on a wintry morning.

Admiral Schuyler writes, "First select a window sill which gets direct sunshine for a couple of hours each day. Inside the window, and well back on the horizontal sill-surface so as to avoid the shadow of the bottom of the window frame when the sun is low in the winter, stick a ⅜-inch-square mirror face upward, using Duco cement. The sun's rays will be reflected upward from this small mirror and, even if they should have to pass through a light rayon curtain, they will make a small, roundish spot of light which travels across the ceiling as the sun's position shifts.

"At some selected time of day, say noon, mark the position of the image on the ceiling with a tack or an inconspicuous cross. Record the date with this mark. Repeat the process at weekly intervals as the seasons change and the sun travels north or south. Now, if you draw a smooth curve on the ceiling through all the marked points, you will be able to tell noon of each day as the moment when the sun's image touches a point on this line. If you also mark its position on the curve on the first day of each month, you can closely estimate the day of the month from the image's crossing on any day.

"Such a curve, labeled 'analemma' and marked with the months of the year, is often reproduced on geographical globes in the conveniently open area of the Pacific Ocean. Our curve on the ceiling is this same analemma—reversed by reflection, distorted by oblique projection on the ceiling, and modified by two sudden displacements when the time changes from 'Standard Time' to 'Daylight Time' and vice versa. But even though the mirror is canted and the ceiling may be somewhat irregular, the method described will establish a time-of-day curve without resort to mathematics or the usual correction devices, required by sundials.

"A single line is perhaps all that many persons will want, but by constructing a series of lines corresponding to the hours and quarter-hour intervals one can readily make a complete sundial. Care must be taken, however, to distinguish the parts of the lines which are traced when the sun is travelling south (June to December) from those when it is travelling north (December to June). One set of markings may be traced by solid lines and the other by dotted lines. . . .

"The travelling spot corresponds to the sun's image in a pinhole camera, the 'pinhole orifice' being replaced by the small mirror. . . . If you watch carefully, cloud formations can often be seen drifting across the sun's disk. Thus the indoor sundial acts as a vane showing the direction of the wind. This effect could doubtless be heightened by inserting a [very weak] spectacle lens of appropriate focal length in the beam—but I have not added this refinement."

SETTING YOUR WATCH BY THE SUN

If your watch stops, you can use it as a sundial and reset it. This trick depends on the fact that the sun moves half as fast as the hour hand. First find south. Then hold the watch flat and point the twelve o'clock mark toward the south. Hold a matchstick vertically on the face of the watch and move it around the numbers on the dial until its shadow falls on the central pivot of the hands. (See Figure 6.) The true time interval before noon or after noon will be just twice the interval indicated by the position of the matchstick on the dial. For

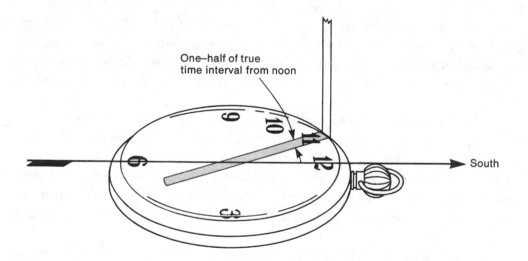

Figure 6. How to set a watch by the sun. With a compass, determine south and point the twelve o'clock in that direction. Hold a small stick upright and move it around the dial until its shadow falls on the central pivot. The time interval from noon, as indicated by considering the stick to be an hour hand, is just half the true interval from noon. In this figure, the true time is two hours before noon, or ten o'clock. (This method is based on the fact that the clock is so constructed that the hour hand moves a full circle in half a day; that is, it moves twice as fast as the sun.)

example, if the stick is at eleven o'clock (one hour from noon), the time is ten o'clock (two hours from noon).

This method will be a bit more accurate if you tilt the watch so its face is perpendicular to the direction of the North Star, which is, of course, invisible.

FINDING SOUTH WITH YOUR WATCH AND THE SUN

If your watch is telling the correct time, you may use it to determine south by a modification of the method described in the preceding section. (Both methods depend on the fact that the hour hand moves 30° per hour while the sun moves 15° per hour.)

Hold a matchstick vertically on the face of the watch at a point just above the tip of the hour hand. Now rotate the watch until the shadow of the match falls on the central pivot of the hands. South will lie halfway between the hour hand and the twelve o'clock mark.

As an example, suppose it is six o'clock in the morning; the sun is due east, and according to the method described the twelve o'clock mark will be due west. South is halfway between.

CONSTELLATIONS AND SIGNS OF THE ZODIAC

Ancient astronomers divided the sun's path about the sky into twelve nearly equal portions and assigned a constellation to each. These are the constellations of the zodiac. ("Zodiac" means "circle of the animals," although not all of the constellations currently represent animals.) On a specific date each year, the sun returns to the same constellation and hides it from our view, so it is easier to determine where the sun is *not* than where it *is*. The following table lists the constellations of the zodiac that are directly south at midnight each month. (These constellations are opposite the sun on these dates, and six months later the sun will be *in* these constellations.)

MONTH	CONSTELLATION IN THE SOUTH AT MIDNIGHT ON THE 21ST
January	Cancer
February	Leo
March	Virgo
April	Libra
May	Scorpius
June	Sagittarius
July	Capricornus
August	Aquarius
September	Pisces
October	Aries
November	Taurus
December	Gemini

The star finder shows the sun's path ("ecliptic") among the stars and indicates the precise locations of the sun at five-day intervals. You will find, for example, that the sun is in Capricornus in January—just six months after Capricornus was directly south at midnight. But if you know something about the astrological "signs" of the zodiac, you will see there is a discrepancy between the constellations and the signs. My birthday, for example, is in January, and I am said to be an "Aquarian" despite the fact that the sun is in Capricornus that month. Astrologers are aware of this discrepancy, and it doesn't seem to disturb them.

The shift between the signs and the constellations is caused by the "precession of the equinoxes," or the shift of the Earth's pole of rotation. Three thousand years ago—and perhaps longer—astronomers chose the sun signs according to the corresponding zodiacal constellations (or perhaps vice versa), and they set Aries at the spring equinox. But the Earth's axis of rotation has shifted since then, and the position of the sun on the first day of spring now lies in Pisces. One entire rotation about the zodiac requires 26,000 years.

THE STAR CLOCK

The dial of the star clock is the zodiac, and the hour hand is your meridian, or north–south line. When the vernal equinox (Pisces) is due south, the star time is zero hours, and each 15° eastward indicates another hour. Taurus, for example, is 60° east of Pisces, and it indicates four hours star time.

The star clock and the sun clock shift against each other as the sun moves eastward among the stars. If you observe the stars at the same sun time, say 9 P.M., you will notice the stars shift slightly westward each night, toward the setting sun. In the course of a year, the stars make one more rotation than the sun, so the star clock rotates more rapidly, by about 1° per day, and the stars set 3 minutes 56 seconds earlier each night than they did the night before. This difference accumulates to one day in a year.

The star clock and the sun clock agree on September 21. At midnight, for example, both clocks read zero hours of the new day. To compute the difference between these clocks at a later date, add 3 minutes 56 seconds for each day that has elapsed, remembering that the star time is *ahead* of sun time.

A precise knowledge of star time is required by those who wish to compute the direction to point their telescope for a particular star, but you will not ordinarily need it. Instead, you may estimate sun time directly from the star clock using the star finder in this book. The method is simple and is described in the next section.

FINDING THE SUN TIME FROM THE STARS
AND THE STAR FINDER

If you know the date, you may find sun time from the stars in the following way. Face south and hold the star finder in front of you with the "south" portion of its horizon downward. Rotate the star disk until it reproduces the observed positions of the stars. A particularly accurate method is to notice the heights of stars that are setting in the west and rising in the east. Then adjust the star finder until these relative heights are correctly reproduced. Now locate the date on the outer scale of the star finder and read the corresponding sun time on the adjacent scale. That's all there is to it.

If you have difficulty aligning yourself toward south, you can turn around and use the stars in the north, especially the Big Dipper, but remember to hold the star finder so the "north" portion of its horizon is down toward the true northern horizon.

DETERMINING THE DATE BY OBSERVING THE STARS AT TWILIGHT

Even if you do not have a watch and do not know the date, all is not lost, because the position of the sun among the stars will tell you the date. Once the date is known, you can use the method of the preceding section to find the time, so the star finder can serve as a calendar as well as a watch.

The trick in determining the date is to find the sun's position among the stars, and twilight is a good time to do this because you can see the stars and you know the sun is just below the horizon.

At the start of morning twilight (about an hour before sunrise) note the positions of the stars, and set the star finder to agree. Then move the star disk ahead one hour to give the positions for the time of sunrise. Now look along the "ecliptic" on the star disk, where the sun's position is marked at five-day intervals, and read the date which lies on the eastern horizon. As an example, if you had been out on the morning of June 22, you would have seen Aldebaran just above the horizon an hour before sunrise. Adjusting the star disk to that position and then moving it ahead an hour to bring the sun to the horizon, you find the date, June 22 (give or take a few days), marked on the zodiac.

Table 1 lists the dates on which other bright stars rise *with* the sun and

TABLE 1

STARS THAT RISE WITH THE SUN ON VARIOUS DATES

STAR OR GROUP	CONSTELLATION	RISES WITH SUN (INVISIBLE)	RISES ONE HOUR BEFORE SUN
Alpha and Beta	Capricornus	Jan. 15	Feb. 7
Alpheratz	Pegasus	Feb. 1	Mar. 1
Circlet	Pisces	Feb. 20	Mar. 20
	Triangulum	Mar. 18	Apr. 16
Pleiades	Taurus	May 16	June 3
Aldebaran	Taurus	June 7	June 22
Betelgeuse	Orion	July 2	July 16
Pollux	Gemini	July 10	July 25
Procyon	Canis Minor	July 27	Aug. 9
Sirius	Canis Major	Aug. 4	Aug. 15
Regulus	Leo	Aug. 21	Sept. 2
Denebola	Leo	Sept. 10	Sept. 22
Arcturus	Boötes	Oct. 7	Oct. 20
Spica	Virgo	Oct. 17	Oct. 29
Alpha	Libra	Nov. 6	Nov. 19
Albireo	Cygnus	Nov. 26	Dec. 6
Antares	Scorpius	Dec. 7	Dec. 18
Altair	Aquila	Dec. 15	Dec. 27

one hour *before* the sun. (Ordinarily, you will not see a star unless it rises almost a full hour before the sun.) Such tables were used by the ancient astronomers to keep track of the seasons.

The date may also be determined by observing the stars after sunset. As before, set the star finder to agree with the observed positions of the stars, but now shift the star disk *back* by the amount of time elapsed since sunset, and read the date where the zodiac crosses the western horizon. (In all of this you must, of course, use the true horizontal direction rather than a horizon obstructed by a mountain.)

RECAPITULATION

In this chapter, the following uses of the star finder have been described:

GIVEN	TO FIND
1. Date	Sun time
2. Nothing	Date

Several variations on these two problems may also be handled, and you may devise your own solutions. For example, given the sun time, you may find the date by noting the position of the stars. (This is Problem 1 in reverse.)

With the star finder you may also predict the times of sunrise and sunset for any day of the year. This is discussed in the next chapter.

Sunset and Sunrise

TIMES OF SUNSET AND SUNRISE

At the spring equinox (March 21) and the fall equinox (September 21) the sun is on the equator and it rises and sets about six o'clock local sun time. In early summer, the sun is in the sky more than twelve hours because its path is then a circle only 67° from the North Pole. The length of the day depends on your latitude and on the date.

Times of sunset and sunrise in a particular city may often be found with the weather forecast in a local newspaper. These times vary with your location, and Table 2, giving times for the *center* of any time zone at various latitudes, may be quite wrong if you are somewhere else, but you may derive corrections for your location by looking outside.

PREDICTING SUNSET AND SUNRISE WITH THE STAR FINDER

For any given date, the times of sunset and sunrise for the center of your time zone and at a latitude of 40° N may be found quite easily with the star finder. (Table 2 may be used as a check.) On the scale of dates running along the zodiac, find the sun's position for the date of interest. Set this point at the eastern horizon (for sunrise) or the western horizon (sunset), and then locate the date on the outer scale of the star disk. Read the time adjacent to the date.

A PECULIAR FEATURE OF SUNSET AND SUNRISE TIMES

The earliest sunset occurs at the end of the first week of December, but the shortest daylight interval occurs two weeks later, at the winter solstice on December 21. The latest sunrise is even later, on the fourth of January. A similar shift occurs at the summer solstice in June. Both are caused by the ellipticity of the Earth's orbit and by the tilt of the Earth's axis of daily rotation from the plane of its orbit. They may be understood the following way.

If the Earth's orbit about the sun were perfectly circular and if the Earth's axis of daily rotation were perpendicular to its orbit, the times of sunset would be the same in all seasons. But the orbit is not circular, and the sun appears to move among the stars more rapidly in winter than in summer, although our clocks run at the *same* rate throughout the year. Thus, our clocks lag behind the sun in summer and move ahead of the sun in winter. This contributes a portion of the peculiarity. The remainder is contributed by the tilt of the Earth's axis, which slows the eastward component of the sun's motion at the time of the equinoxes. As a result, the true position of the sun can deviate by as much as

TABLE 2
TIMES OF SUNRISE AND SUNSET

| | | SUNRISE | | | SUNSET | | |
		35°N	40°N	45°N	35°N	40°N	45°N
LATITUDE							
DATE							
Dec.	31	7:08	7:22	7:38	16:58	16:44	16:28
Jan.	15	7:08	7:20	7:35	17:12	16:59	16:44
	30	7:01	7:11	7:23	17:27	17:17	17:05
Feb.	14	6:48	6:55	7:03	17:42	17:34	17:26
Mar.	1	6:30	6:34	6:39	17:56	17:52	17:47
	16	6:10	6:11	6:11	18:08	18:08	18:07
	31	5:49	5:46	5:43	18:20	18:23	18:26
Apr.	15	5:28	5:23	5:16	18:32	18:38	18:45
	30	5:11	5:02	4:51	18:44	18:53	19:04
May	15	4:57	4:45	4:31	18:56	19:08	19:22
	30	4:48	4:34	4:18	19:07	19:21	19:38
June	14	4:45	4:30	4:13	19:15	19:30	19:48
	29	4:49	4:34	4:16	19:18	19:33	19:51
July	14	4:56	4:43	4:26	19:15	19:29	19:45
	29	5:07	4:55	4:41	19:06	19:17	19:31
Aug.	13	5:18	5:09	4:59	18:51	19:00	19:10
	28	5:29	5:24	5:17	18:33	18:38	18:45
Sept.	12	5:40	5:38	5:35	18:12	18:14	18:17
	27	5:51	5:52	5:53	17:50	17:49	17:49
Oct.	12	6:03	6:07	6:11	17:29	17:26	17:21
	27	6:16	6:23	6:31	17:11	17:04	16:56
Nov.	11	6:30	6:40	6:52	16:57	16:48	16:36
	26	6:45	6:57	7:12	16:50	16:37	16:23
Dec.	11	6:58	7:12	7:28	16:49	16:35	16:18
	26	7:06	7:20	7:37	16:55	16:41	16:24

sixteen minutes from the position of the fictitious, average sun to which our clocks are adjusted. This deviation, known as the "equation of time," produces the peculiarity of sunset and sunrise times.

GUESSING THE TIME REMAINING UNTIL SUNRISE AND SUNSET

Morning twilight begins when the sun is 12° to 15° below the horizon and its rays strike the upper layers of the atmosphere. The interval between this moment and the time of actual sunrise will vary with the season and your latitude, but one hour is a useful estimate for the middle latitudes.

Even without a watch, it is possible to make a surprisingly good guess of the time remaining until sunset—on a clear day. If you spread your thumb and fingers, and then extend your arm, the distance between the tip of your thumb and the tip of your index finger will be approximately 15°, the sun's motion in one hour. (Not all hands are the same, so you should develop a technique that makes your hand extend one twenty-fourth of a full circle.) Using your extended hand, measure the distance from the sun to the point on the horizon where the sun will set. Doing this, you can guess the corresponding time interval to within a quarter-hour with a little practice.

TWILIGHT COLORS

As far as colors go, sunrise is just sunset in reverse, so I will not distinguish between them. (Also, moonrise and moonset would appear just as colorful as sunset if our eyes were more sensitive to faint light.)

Sunset colors fade when the sun has descended too far to illuminate the atmospheric clouds and dust, so the duration of these added colors is determined by two factors: the varying height of the sun and the height of clouds and dust in the air.

White light is a mixture of all colors, but the light of sunset has been scattered and absorbed by clouds, dust, smoke, and air. The setting sun is red because its blue light is cut to one five-millionth and the red light to one two-thousandth of its original strength. Clouds catch this red light without altering its color, because their droplets are large enough to reflect the light of all colors equally.

Look eastward immediately after sunset and you will see the horizon darken as the broad, indistinct shadow of the Earth climbs rapidly up the sky, leaving the clouds and lower air in darkness. (This is the only time of the day that the horizon is darker than the sky overhead. At all other times—even at night—the horizon is brighter than the sky overhead.)

Thirty minutes later, you will see a pink glow in the western sky half-way toward the zenith. This is sunlight scattered by dust in a narrow layer at a height of forty miles.

THE SHAPE OF THE SETTING SUN

The setting sun is highly flattened because the Earth's atmosphere raises the lower portion of the sun more than the upper portion. This *refraction* is sufficient to keep the sun above the horizon two minutes after it would have set. (On Venus, refraction would carry sunlight all around the planet. If it were not for clouds, the sun would appear never to set; it would appear to spread all around the horizon.)

THE MIDNIGHT SUN

Those who live north of the Arctic Circle (latitude 66½° N) see the midnight sun during a portion of the summer. The farther north, the longer is the interval; there is a corresponding interval without sunlight during winter. The following table lists the intervals of the midnight sun for various latitudes.

LATITUDE (N)	INTERVALS OF MIDNIGHT SUN
67°	June 6–July 9
71°	May 15–July 31
75°	Apr. 30–Aug. 14
79°	Apr. 18–Aug. 27
83°	Apr. 7–Sept. 7

Eclipses

LUNAR ECLIPSES

With luck, you may see several eclipses of the moon each year, no matter where you are. These eclipses occur when the moon enters the Earth's shadow, which consists of two cones pointing away from the sun. One, the narrow *umbra,* converges and vanishes at a distance of one million miles, well beyond the moon. The other, the broad *penumbra,* diverges and extends endlessly. Within the umbra, the sun's face is completely hidden from the moon; within the penumbra, it is partially hidden. See Figure 7.

A lunar eclipse commences with the passage of the Earth's faint and indis-

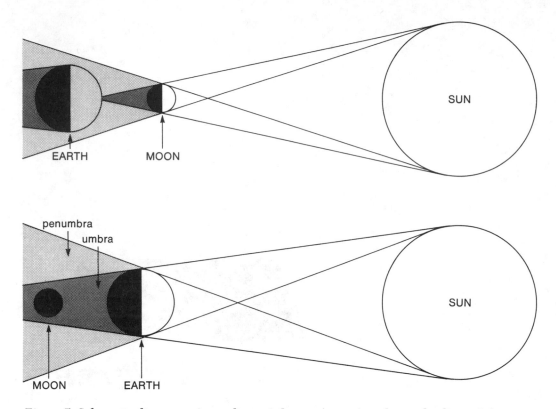

Figure 7. Schematic diagrams of an eclipse of the sun (upper) and a total eclipse of the moon (lower). The Earth's shadow (and the moon's) consists of two portions, the dark inner umbra and the penumbra. A traveler in the penumbral region would see the sun partially eclipsed; in the umbra the sun appears totally eclipsed.

tinct penumbral shadow over the face of the moon. This phase of the eclipse is difficult to see, but after an hour or so the edge of the umbral shadow strikes the moon's face. The umbral shadow is dark and slightly reddish; the penumbral shadow is bright and silvery by comparison. At totality the moon will nearly vanish, but in most eclipses a remnant of light remains—light that has passed through the atmosphere of the Earth. (During several recent eclipses, astronomers have used world-wide cloud data to show that clouds in our atmosphere affect the brightness of the eclipsed moon.)

SOLAR ECLIPSES

The moon also casts umbral and penumbral shadows. During a partial eclipse of the sun, we stand in the penumbral shadow of the moon. The umbral shadow of the moon is about one-quarter as long as the Earth's, because the moon's diameter is one-quarter the Earth's, and the most amazing coincidence of the solar system is that the moon's umbral shadow reaches precisely to the Earth's surface. To express this coincidence in slightly different terms, the moon has precisely the correct size to cover the face of the sun as seen from the Earth. (And, to say it in a third way, the moon's distance is 1/400 the sun's distance and its true size is 1/400 the sun's true size.)

As a result of this coincidence, total solar eclipses are spectacular and they may be seen only from a very narrow band of the Earth's surface.

The sight of a total eclipse of the sun will repay a few hundred miles' travel, even for the person who does nothing but gawk. The eclipsed sun is a black hole in the sky, surrounded by the glowing corona of the sun's atmosphere. Stars appear, and the horizon seems engulfed by an immense forest fire. For me, the most amazing feature of an eclipse is its silence. I expect the grinding of machinery or a few claps of thunder, but I only hear the barking of dogs and the evening songs of confused birds. Until you have seen a total eclipse, it is difficult to imagine how it differs from a partial eclipse and from an eclipse of the moon.

WHERE TO GO TO SEE AN ECLIPSE

Tracks of the sun's umbral shadow during the next few years are shown in Figure 8. For more details, consult *Sky and Telescope* magazine* a year or so before the eclipse.

---

* This magazine, published in Cambridge, Massachusetts, is available in many public libraries.

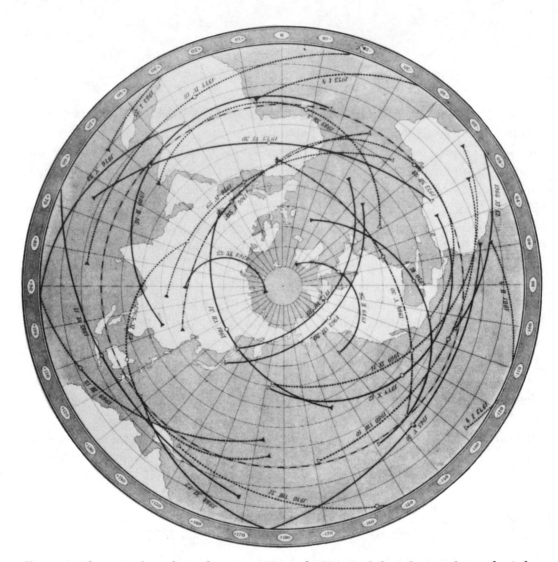

Figure 8. Chart of solar eclipses between 1963 and 1984. Each line depicts the track of the central point of the moon's shadow for the year, month, and day indicated. In each case, the moon's shadow strikes the Earth at the western end of the track, moves eastward at about 1000 miles per hour, and then lifts from the surface. Total eclipses are indicated with solid lines, annular eclipses with broken lines. Forthcoming lunar and solar eclipses are listed at the end of this book. Lunar eclipses, being visible from the entire night hemisphere, require no chart. (Reproduced by permission from Canon of Eclipses, *T. R. von Oppolzer, trans. Owen Gingerich, New York, 1962.)*

THE SEASONS OF ECLIPSES

If the moon followed precisely the sun's path around the sky, we would see a solar eclipse every month at new moon and a lunar eclipse at full moon, but the 5° tilt of the moon's orbit prevents this, as is indicated in Figure 9. The moon's monthly track crosses the sun's path at two points called the *nodes,* and an eclipse can occur only when the sun is near one of them. Because the sun requires six months to move from one node to the other, eclipses occur at intervals of about six months. In 1976, for example, eclipses will occur in April and October. During the following years, eclipses will occur earlier, and the dates are given at the end of the book.

Figure 9. Diagram illustrating why eclipses only occur at two seasons of the year. The moon's orbit is tilted 5° from the Earth's orbit, and the two orbital planes intersect along the "line of nodes." Eclipses of the sun and moon can only occur when the Earth's orbital motion carries it to the places where the line of nodes points directly at the sun. These two places are on opposite sides of the orbit, so eclipses occur at two times a year, six months apart. (A slow movement of the line of nodes causes the eclipses to occur at slightly different times each year.)

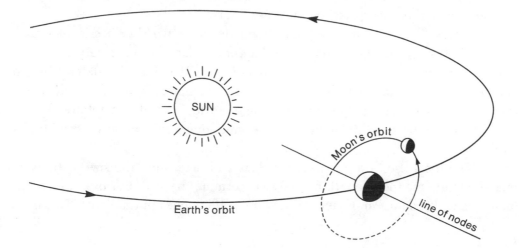

The moon's orbit rotates around the ecliptic, and the nodes move westward, making a complete cycle in eighteen years. Thus, eclipses occur several weeks earlier in succeeding years. (The eighteen-year cycle is called the Saros, and it was known to the ancients.)

ECLIPSE PHOTOGRAPHY

A great deal happens in a short time during a solar eclipse, and if you make a photographic record you will probably see many things you missed during the actual eclipse. (Of course, you also run the danger of having to fuss with your camera at the height of the eclipse when everyone else is calmly admiring the spectacle.) I find color transparency film far more satisfactory than black and white, particularly when the pictures are projected onto a large screen. And movies are fascinating.

During the few minutes before and after totality, the sun's crescent appears deceptively faint and harmless, but it is nearly as dangerous as the full sun. Thus it is not difficult to burn a hole in your retina by looking at the partially eclipsed sun, either with the naked eye or through the viewfinder of your camera. Also, the heat of the sunlight can burn a hole in your camera shutter or cause it to warp, so you must use filters.

A safe filter for your eyes may be made from two thicknesses of heavily exposed and fully developed photographic film. Polaroids and gelatin filters are not safe, because they transmit infrared radiation that may be damaging. For your camera, a No. 5 neutral-density filter is adequate. These filters should be removed during totality. There is no danger in looking at the totally eclipsed sun, but don't be caught looking at the emerging crescent after totality has passed.

The focal length of your camera lens will determine the size of the sun's image. Divide the focal length by 100 to determine the size. For example a four-inch lens will produce an image on the film 4/100 inches in diameter. This is the

TABLE 3
SOLAR ECLIPSE PHOTOGRAPHY
APPROXIMATE EXPOSURES FOR STILL AND MOVIE CAMERAS

ASA SPEED		PARTIAL PHASES		TOTALITY (PROMINENCES)		TOTALITY (INNER CORONA)		TOTALITY (OUTER CORONA)	
		STILL	MOVIE	STILL	MOVIE	STILL	MOVIE	STILL	MOVIE
25–32	Lens Opening	f/5.6	f/11	f/4.5	f/8	f/4.5	f/2.8	f/4.5	f/1.4
	ND Filter	5.00	5.00	none		none		none	
	Time (seconds)	1/100	16 FPS	1/100	16 FPS	1/10	16 FPS	1/2	16 FPS
40–50	Lens Opening	f/6.3	f/13	f/5.6	f/11	f/5.6	f/3.5	f/5.6	f/1.9
	ND Filter	5.00	5.00	none		none		none	
	Time (seconds)	1/100	16 FPS	1/100	16 FPS	1/10	16 FPS	1/2	16 FPS
64	Lens Opening	f/8	f/16	f/6.3	f/13	f/6.3	f/4	f/6.3	f/2
	ND Filter	5.00	5.00	none		none		none	
	Time (seconds)	1/100	16 FPS	1/100	16 FPS	1/10	16 FPS	1/2	16 FPS
125–160	Lens Opening	f/11	f/22	f/8	f/16	f/8	f/4.5	f/8	f/2
	ND Filter	5.00	5.00	none		none		none	
	Time (seconds)	1/100	16 FPS	1/100	16 FPS	1/10	16 FPS	1/2	16 FPS
200–250	Lens Opening	f/16	f/9.5	f/11	f/22	f/11	f/6.3	f/11	f/2.8
	ND Filter	5.00	6.00	none		none		none	
	Time (seconds)	1/100	16 FPS	1/100	16 FPS	1/10	16 FPS	1/2	16 FPS
400–650	Lens Opening	f/22	f/11	f/16	f/11	f/16	f/9.5	f/16	f/4
	ND Filter	5.00	6.00	none	1.00	none		none	
	Time (seconds)	1/100	16 FPS	1/100	16 FPS	1/10	16 FPS	1/2	16 FPS
1250	Lens Opening	f/32	f/16	f/22	f/16	f/22	f/13	f/22	f/5.6
	ND Filter	5.00	6.00	none	1.00	none		none	
	Time (seconds)	1/100	16 FPS	1/100	16 FPS	1/10	16 FPS	1/2	16 FPS

FPS = Frames Per Second.

thickness of a dime. When a transparency is projected onto a screen with a magnification of 100×, the resulting image of the sun will have a diameter equal to the focal length of the camera.

During the partial phases, the sun should be photographed with the longest focal length available, and you should also notice the peculiar crescent-shaped shadows produced by light filtering through the leaves of trees.

During totality, a short focal length lens can capture an interesting general view of the eclipse and will reveal the glowing sky beyond the limits of the moon's shadow. With a moderate telephoto lens you can catch the general shape of the sun's corona, a pale glowing atmosphere that extends many times the diameter of the sun. With an extreme telephoto, you will see details of the coronal streamers, and you may be able to catch Baily's beads, spots of sunlight shining through valleys on the moon during the few seconds before and after totality.

Vary your exposures and shoot lots of film. It is difficult to make precise predictions of the best exposure for each set of conditions, but the guidelines in Table 3, developed by the Eastman Kodak Company, give representative values based on the experience of numerous photographers. (Further information can be found in *Solar Eclipse Photography for the Amateur,* pamphlet AM-10, available from Consumer Markets Division, Eastman Kodak Company, Rochester, N.Y. 14650. See also *Sky and Telescope* magazine, June, 1972, page 358.)

Catching
the Unpredictable

SUNSPOTS

If you use the following method, sunspots are quite easy to see. Focus a pair of binoculars or a telescope on an object about forty feet away. Then support the binoculars against a railing or a tree trunk, place a sheet of white paper two feet behind the eyepiece, and point the binoculars toward the sun. (For heaven's sake, DON'T LOOK THROUGH THEM.) By adjusting the focus and moving the paper toward and away from the eyepiece, you will be able to throw a sharp image of the sun onto the paper. An image two inches across will make the sunspots fairly easy to see—if there are any.

Many sunspots are larger than the Earth, but with binoculars they look like fly specks. Move the paper and keep looking. You will probably find two or three.

Sunspots usually persist for at least a week, and if you watch for a few days you will be able to see them move with the rotation of the sun. In two weeks a spot will move entirely across the sun's face.

Sunspots are areas which have been cooled by magnetic fields in the sun's atmosphere, and they are also the location of the fierce magnetic storms— "flares"—that shoot solar atoms toward the Earth and cause the northern lights.

NORTHERN LIGHTS (AURORAE)

The high layers of the atmosphere often glow brightly as atoms from the sun enter the Earth's magnetic field.

Aurorae are not connected with the seasons, but they are more often seen during the winter because the nights are longer and the air tends to be less hazy. Bright ones show patches of red or green, and sometimes both. These colors, like those of an electric sign, are produced by colliding atoms. Pulsations of brightness are not unusual.

Aurorae are rarely seen near the equator. The region of greatest activity is a lopsided ring passing over Canada, Alaska, Siberia, and Scandinavia. In this zone, aurorae can be seen on almost every clear, moonless night of the year.

WHERE TO HUNT THE RAINBOW

Raindrops and sunlight can also produce colors when they are mixed in the right part of the sky. (It is not necessary to have raindrops falling on your head.) Actually there are two bows of color, and the strongest forms a circle of radius 42° around the direction opposite the sun. The weaker bow lies 9° outside, and its colors are reversed. (This reversal occurs because the sunlight bounces twice inside the drop rather than once, which also accounts for the faintness of the secondary bow, because a great deal of light is lost on each bounce.)

You must always turn your back on the sun to see a rainbow. If the sun is nearly setting, the center of the bow will lie just below the opposite horizon, and the bow will stand high in the sky. When the sun is high overhead, you must look farther down to see the rainbow, and you may see one in a garden sprinkler or the spray of a boat. (Rainbows should also be visible from the top of a mountain when the sun is high.)

HALOES AND SUNDOGS

Ice crystals floating in the air can also produce circular arcs of light, although their colors are much less striking than those of the rainbow. At night, a circular halo about the moon reveals a thin layer of cirrus clouds, and it often portends a change for the worse in the weather.

Many other forms of haloes may be seen in sunlight. A halo of 22° radius centered on the sun is not uncommon, and in late afternoon it is often possible to see diffuse spots of faintly colored light on either side of the sun at a distance of 22°. These are the "sundogs." And there are other, less common forms produced by ice crystals of various shapes falling through the air in various positions. Although they are indistinct, these arcs and haloes may be photographed. Color film is preferable.

COMETS

Every few years a comet passes near the sun and becomes visible to the naked eye for several weeks. The head of a comet is composed of a "snowball" a few miles in diameter surrounded by a diffuse halo. The tail of a comet is composed of gas and dust released from the nucleus by the heat of sunlight and propelled outward by sunlight and solar atoms. Some comets have several tails, each ejected from a different portion of the nucleus or composed of different types of particles.

Many of the brighter comets have very long orbits, requiring hundreds of years—or thousands. Halley's Comet is a magnificent exception with a period of only seventy-five years. It will be seen in 1986, and its tail will reach halfway across the sky, so there will be no missing it. Most comets are not as easy to spot, and a good pair of binoculars is a great help. The sky must be very clear.

There is no telling when a new, bright comet will appear, but of the known comets, Encke's and Tuttle's may become visible with binoculars during December, 1980.

Photography of comets that are visible to the naked eye is not difficult, and the procedure is described in the February, 1976, issue of *Astronomy* (published by AstroMedia Corporation, 757 N. Broadway, Milwaukee, Wisconsin 53202). Use fast film, mount the camera on a tripod, open its lens all the way, and try exposures of 10 to 60 seconds. Longer exposures will give trailed images. (This technique can also be used to photograph constellations.)

METEORS

The flash of a meteor is caused by a fragment, often no larger than a fingernail, descending through the atmosphere at speeds of ten to fifty miles per second. We see them at heights of about fifty miles, and they seldom reach the ground.

Meteors come in two varieties: "sporadic" and "shower." On every dark night of the year, it is possible to see a sporadic, or random, meteor every few minutes. There is no telling where the next will appear, because they are scattered through the solar system and presumably are fragments left over from the formation of the planets. Occasionally a very large fragment will produce a "fireball," remaining visible for hundreds of miles and crashing to earth as a meteorite.

Meteor "showers" are produced by swarms of cometary debris that · follow their parents about the sun. Every comet produces such a swarm, but we encounter only a few of them. When we do, the meteors appear to fan out from a single point in the sky. This is the "radiant," and the shower is named for the constellation or star near that point. On the night of a shower, hundreds of meteors may be seen in a few hours. Table 4 lists the best showers. They appear on the same date each year, because their orbits are fixed in the solar system.

Little is gained by looking for meteors with binoculars or telescopes because such devices restrict your field of view. Photography is not difficult. Leave the camera open for a half-hour or so. Moonlight is a definite hindrance (and will make such photography impossible), so it is best to look for meteors in a moonless sky.

TABLE 4
THE MOST INTENSE METEOR SHOWERS

NAME	DATES	MAXIMUM HOURLY RATE	ASSOCIATED COMET
Quadrantids	Jan. 3	50	?
Eta Aquarids	May 5±4	20	Halley
Delta Aquarids	July 28±3	20	?
Perseids	Aug. 12±2	50	1862 III
Orionids	Oct. 21±3	20	Halley
Geminids	Dec. 12±2	50	?

Some Odds and Ends

THE TOTAL NUMBER OF VISIBLE STARS

On the entire celestial sphere there are about 5000 stars brighter than sixth magnitude, the limit for average eyesight, and in an open space on a clear night you can see approximately 2000. The visible stars give only one-fifth of the total light of the stars; the remainder is contributed by stars too faint to be seen individually.

And starlight itself contributes less than one-sixth of the total light of the sky. Even on the darkest night, faint aurorae and light scattered by dust among the planets give five times more light than the stars.

Binoculars with two-inch (50 mm.) lenses will permit you to see about 235,000 stars on the celestial sphere—nearly fifty times as many as the naked eye. And with a six-inch telescope, you would be able to count 2,600,000. The largest telescope in the world would reveal more than a billion stars.

THE CENTER OF OUR GALAXY

Our galaxy contains two dozen stars for every human, or about 100 billion. Only one percent can be photographed individually even with the largest telescopes; the remainder produce the diffuse band of the Milky Way. Many stars in our galaxy are hidden by clouds of interstellar dust, and these clouds give the Milky Way a ragged and irregular appearance.

The brightest and broadest portion of the Milky Way lies in Sagittarius, and it may be seen low in the south during summer evenings. This is the center of our galaxy.

Determining the precise size of our galaxy has been a difficult task, and one of the best methods is the one used by Harlow Shapley in 1916. He measured brightnesses of immense clusters of stars swarming about the center, and when corrections are made for the obscuration by dust, the result is 25,000 light-years.

How can the total number of stars in our galaxy be estimated if 99 percent of the stars are hidden by dust? The answer is a guess based on the speed with which our galaxy rotates, and the method is similar to that used for determining the mass of the sun from the speeds and distances of the planets in the solar system. Both types of motion are governed by Newton's laws. Once the distance of the center is known and the speed of rotation has been measured, the mass of the central body may be computed. The sun moves around the center of our galaxy at 150 miles per second, and this figure gives a total mass of about 100 billion suns for our galaxy. This puts us in the upper ranks of galaxies, but several of our neighbors are larger.

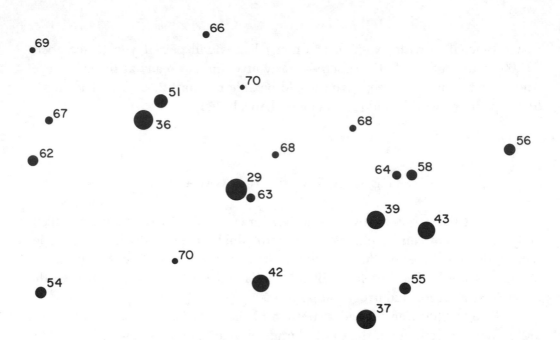

Figure 10. Stars of the Pleiades cluster ("The Seven Sisters") brighter than seventh magnitude. This diagram will permit you to test the visibility of faint stars. Each star is labeled with its magnitude in visual light, the decimal point being omitted. (Thus, "29" indicates magnitude 2.9.) On a dark night with a clear sky, the average eye can see stars to the sixth magnitude. This cluster lies in the constellation Taurus, and it is indicated on the star finder.

THE MAGNITUDES AND BRIGHTNESSES OF STARS

The ancients divided the visible stars into six steps of brightness, called *magnitudes*. The twenty brightest stars were placed in the first magnitude, and the faintest were placed in the sixth. The steps were uniform, so stars of second magnitude, for example, appeared to be halfway between those of the first and third.

Experimenters of the nineteenth century discovered that these magnitude steps correspond to equal *ratios* of brightness, and the ratio between the ancient classes of first and second magnitude is very nearly the same as the ratio between the brightnesses of second- and third-magnitude stars, and so on. This rule is the basis for the modern definition of stellar magnitudes, which states that a difference of one magnitude corresponds to a brightness ratio of 2.512. This number was chosen so a difference of five magnitudes would correspond precisely to a ratio of 100 in brightness. Table 5 gives the relative brightnesses of stars of different magnitudes, and Figure 10 indicates the magnitudes of the brighter stars of the Pleiades.

TABLE 5

STELLAR MAGNITUDES AND BRIGHTNESSES RELATIVE TO THE FAINTEST NAKED-EYE STARS

MAGNITUDE	RELATIVE BRIGHTNESS	
−1	631	
0	251	
1	100	
2	39.8	
3	14.8	
4	6.31	
5	2.51	
6	1.00	Faintest Naked-Eye Stars
7	0.398	
8	0.158	

Although stars of the sixth magnitude are the faintest that can normally be seen with the naked eye on a dark night, the magnitudes can be extended further with telescopes and cameras. (An eleventh-magnitude star is 100 times fainter than a sixth-magnitude star, and a sixteenth-magnitude star is fainter by another factor 100. The faintest stars that have been photographed are 4 million times fainter than the limit of the naked eye.)

Table 6 gives the total number of stars brighter than each magnitude. For example, of all the stars in the sky, 150 are brighter than third magnitude.

TABLE 6
COUNTS OF THE STARS BY MAGNITUDE

MAGNITUDE	NUMBER BRIGHTER	MAGNITUDE	NUMBER BRIGHTER
−2.0	0	5.0	1,700
−1.0	1	6.0	4,900
0.0	4	7.0	14,000
1.0	16	10.0	350,000
2.0	45		
3.0	150	15.0	37,000,000
4.0	540	20.0	1,200,000,000

THE LIMIT OF FAINTNESS IN A TELESCOPE

The larger the area of a telescope lens or mirror, the fainter the stars that may be seen. The first two columns of Table 7 list the *limiting*, or faintest, magnitude that can be seen through telescopes of different diameters.

Also, when a photographic film is placed at the focus of a telescope, transforming it into a camera, stars can be detected nearly 100 times fainter than could be seen through the same instrument. Table 7 summarizes the limiting magnitudes that can be photographed with telescopes of various sizes. (This table is merely representative, because not all types of telescopes will have the same limits, nor is there universal agreement among astronomers about the practical limit of a given telescope.)

TABLE 7
LIMITING MAGNITUDES FOR REFLECTING TELESCOPES

TELESCOPE DIAMETER (INCHES)	LIMITING MAGNITUDE* CONSISTENTLY SEEN	TOTAL STARS VISIBLE ON ENTIRE SPHERE	PHOTOGRAPHIC LIMIT (f/5 MIRROR)**
Eye	6.0	5,000	
2	9.7	235,000	15.0
3	10.6	600,000	15.5
4	11.3	1,200,000	16.2
5	11.8	1,900,000	16.8
6	12.2	2,600,000	17.5
8	12.9	5,250,000	18.0
10	13.4	7,500,000	18.3
12	13.8	12,000,000	18.5
24	15.5	52,000,000	19.8
48	17.0	190,000,000	21.0
100	18.8	600,000,000	22.0
200	20.3	1,300,000,000	23.0

* A difference of 5 magnitudes gives a brightness ratio of 100.
** The limit of an f/3 mirror is one magnitude brighter.

COLORED STARS

Sunlight is yellowish white, and this is typical of most stars that we can see, but there are some whose color is distinctly red or blue. (I have never seen a star that looked green, but others say they have.)

The eye has difficulty with the colors of faint stars, so you should concentrate on the brightest ones, and a pair of binoculars is helpful. The tints are as vivid as the colors in a flame.

With the star finder you can locate the following stars whose colors are noticeable. Each is among the brightest in its constellation.

TABLE 8
STAR COLORS

NAME	CONSTELLATION	COLOR
Spica	Virgo	Blue
Arcturus	Boötes	Yellow
Pollux	Gemini	Yellow
Castor	Gemini	White
Capella	Auriga	Yellow-white (sunlight)
Rigel	Orion	Blue-white
Antares	Scorpius	Red

TESTING YOUR EYES AND BINOCULARS ON DOUBLE STARS

A test of eyesight reportedly used by the Arabs is the pair of stars Mizar and Alcor, in the handle of the Big Dipper. Mizar is as bright as the North Star (magnitude 2) and its companion is one-fourth as bright. They are separated by eleven minutes of arc. (One minute of arc is one-sixtieth of a degree. The diameters of the sun and moon are thirty minutes of arc.) Most people can separate Mizar and Alcor quite easily with their naked eyes. (Mizar itself is a double star, but the separation is only 14.5 seconds of arc, so a telescope is required.)

In the summer sky, the star α Capricornus* provides a good test of eyesight. Its components are separated by six minutes of arc. This star forms the upper right-hand tip of the crescent-shaped constellation Capricornus.

Two "stars" in Scorpius may also be seen as double with the naked eye on a clear summer night. They are listed in Table 9 and are indicated on the star finder with the symbol -•-.

TABLE 9
NAKED-EYE DOUBLE STARS

DESIGNATION	COORDINATES	MAGNITUDES		SEPARATION (ARC MINUTES)
		BRIGHTER	FAINTER	
Mizar and Alcor	13ʰ22ᵐ +55°11′	2.17	4.02	11
ι Scorpii	17 44 −40 06	3.14	4.88	9
μ Scorpii	16 48 −37 57	3.09	3.64	7
α Capricornus	20 15 −12 40	3.77	4.55	6

Even the best of eyes can rarely distinguish two stars closer than three minutes of arc, so the star ε Lyrae (separation 3.5 minutes) provides a tough test. This star is near the bright star Vega (see star finder), and my sons say they can distinguish its components quite easily, although I cannot. In a five-inch telescope this star is quite a remarkable sight, because each component can be seen as a double, and this is a *quadruple* system.

Rigidly mounted binoculars will reveal many other pairs of stars, and you will probably be able to distinguish stars whose separations are greater than five

---

* "Alpha" is the brightest star in Capricornus. Greek letters have been assigned to the brighter stars, more or less alphabetically in order of decreasing brightness. The Greek alphabet is listed at the end of this book.

minutes of arc *divided by* the magnifying power of the binoculars. In seconds of arc, this rule gives 300/(magnifying power), or about 45 seconds for 7× binoculars. If you merely hold the binoculars in your hands instead of mounting them, your limit will probably be about half as sharp, judging from my own experience.

Table 10 lists bright double stars whose components are separated by more than forty seconds of arc and have nearly equal brightness. These stars are indicated on the star finder.

Table 11 lists double stars that are beyond the reach of binoculars, but that may be separated with a small telescope.

Some of the double stars in these lists are accidental pairings of stars lying at vastly different distances from the sun. However, most of the fainter and more difficult doubles are physically connected and are moving in gravitational orbits about one another, much as the planets move around the sun. In fact, the majority of stars are members of multiple systems. (No planets have been detected beyond the solar system, but several stars appear to be twirled by a dark companion.)

TABLE 10
DOUBLE STARS VISIBLE WITH BINOCULARS

DESIGNATION	COORDINATES		COMBINED	MAGNITUDES BRIGHTER	FAINTER	SEPARATION (ARC SECONDS)	NAME
α Librae	14ʰ48ᵐ	−15°50′	2.79	2.90	5.33	231	*Zubenelgenubi*
ε Lyrae	18 43	+39 37	3.83	4.50	4.68	208	
α Leonis	10 05	+12 13	1.34	1.34	7.64	176	*Regulus*
μ Boötis	15 22	+37 33	4.33	4.47	6.67	109	*Alkalurops*
16, 17 Dra	16 35	+53 01	4.65	5.20	5.64	91	
υ Draconis	17 31	+55 13	4.20	4.95	4.98	62	
ζ Lyrae	18 43	+37 32	4.06	4.29	5.87	44	

TABLE 11
DOUBLE STARS VISIBLE IN SMALL TELESCOPES

| | COORDINATES | | MAGNITUDES | | | SEPARATION (ARC SECONDS) | |
DESIGNATION	α	δ	COMBINED	BRIGHTER	FAINTER		NAME
υ Scorpii	16ʰ09ᵐ	−19°20′	4.16	4.29	6.49	41	*Lesath*
β Cygni	19 29	+27 52	3.10	3.24	5.36	35	*Albireo*
ι Cancri	8 43	+28 57	4.09	4.20	6.61	31	
ψ Draconis	17 43	+72 11	4.58	4.90	6.07	30	
θ Serpentis	18 54	+ 4 08	4.10	4.50	5.37	23	*Alya*
ϵ Monocerotis	6 21	+ 4 37	4.33	4.48	6.54	13	
α Can Ven	12 54	+38 35	2.80	2.90	5.39	20	*Cor Caroli*
ζ Ursa Majoris	13 22	+55 11	2.17	2.40	3.96	14	
κ Boötis	14 12	+52 01	4.44	4.60	6.61	13	
β Scorpii	16 02	−19 40	2.76	2.90	5.06	13	*Acrab*
γ Delphini	20 44	+15 57	4.12	4.49	5.47	10	

VARIABLE STARS

Several prominent stars show variations of brightness that are easy to spot if you compare these stars with their neighbors.

A good example in the summer evening sky is β Lyrae. This "star" is actually two stars in rotation around a common center with a period of two weeks. When the fainter star eclipses the brighter, β Lyrae fades to one-half its usual brightness and remains faint for about one day. You can see it as the star in the corner of the parallelogram of Lyra farthest from the bright star Vega. When at its brightest, β Lyrae equals its companion at the end of the parallelogram, γ Lyrae, but on some nights it will be as faint as the star in the far corner, δ Lyrae.

An eclipsing star in the autumn and winter sky is Algol, or β Persei. Its period is three days, and it fades to one-third its normal brightness during the eclipses, which last for several hours.

In addition to eclipsing stars, there are several other classes. In the winter evening sky, one star to watch is Mira, the "Wonderful." It is in the constellation Cetus, the Whale, and it reaches third or fourth magnitude at brightest, then fades to ninth magnitude about once a year. It totally vanishes for six months at a time, and its behavior is quite erratic. Mira belongs to the second class of variable stars, the *pulsators*. Its surface brightness increases and decreases as heat from the nuclear burning in the center is alternately blocked and released by the body of the star. Its surface layers also move in and out, changing the total area of the star. Pulsators represent a short stage in the life of a typical star, a stage when the star cannot remain still—somewhat like the flame of a campfire.

The third class of variable star, the *nova* or exploding star, has been very important for measuring the distances to other galaxies, because novae become 10,000 times brighter than the sun and can be seen at great distances. Many novae have occurred in our galaxy, and their explosions are evidently caused by the exchange of matter between two close stars; one star becomes overloaded and its surface layers erupt like a nuclear bomb. Despite the violence of the event, the stars are not destroyed.

Supernovae are the fourth class of variable star, and they become thousands of times brighter than an ordinary nova. Again, the cause is a bomblike explosion, but in these cases, the bulk of star is scattered in space, leaving a small dense residue: a neutron star. Supernovae are relatively rare; one occurs every hundred years or so in our Milky Way and at a similar rate in other galaxies like our own.

Table 12 lists the variable stars that become bright enough to be seen with the naked eye. These stars are indicated on the star finder with the symbol ⊙.

The regular observing of and reporting on variable stars is a hobby of thousands of amateurs, whose work is of direct value to professional astronomers.

The simplest equipment will suffice, and information on techniques, charts, and reports can be obtained from The American Association of Variable Star Observers, 187 Concord Avenue, Cambridge, Massachusetts 02138.

TABLE 12
BRIGHT VARIABLE STARS

NAME	POSITION α	δ	MAGNITUDE BRIGHTEST	FAINTEST	PERIOD (DAYS)	TYPE
γ Cassiopeiae	0ʰ54ᵐ	+60°27′	1.6	2.9	Irregular	NI
o Ceti	2 17	− 3 12	3.4	9.2	331	P
β Persei	3 05	+40 46	2.4	3.5	2.87	E
ε Aurigae	4 59	+43 45	3.3	4.1	9900	E
α Orionis	5 52	+ 7 24	0.5	1.1	2070	P
β Lyrae	18 48	+33 18	3.4	4.1	12.91	E
η Aquilae	19 50	+00 52	3.8	4.5	7.18	P
μ Cephei	21 42	+58 33	3.7	4.7	Irregular	P
δ Cephei	22 27	+58 10	3.6	4.2	5.37	P

P = Pulsator; E = Eclipsing binary; NI = Intermittent explosive activity.

THE ORION NEBULA

Hanging below the "belt" of Orion is a chain of three rather indistinct "stars." The central "star" appears to be fuzzy, and it is actually a small cluster of five blue stars. With large binoculars, these stars are seen to be surrounded by a pale cloud of glowing gas. This is the Orion nebula, and the spectroscope proves its light to originate in a gas excited by the light of nearby stars. It lies within our Milky Way.

Photographs of the Orion nebula show dozens of small dark spots produced by dense clouds of dust. These clouds may contract and become stars, but a million years will have elapsed before they have become hot enough to shine like the sun.

THE ZODIACAL LIGHT

This is a very faint band of light (roughly as wide as the Milky Way) which may be seen standing above the horizon among the constellations of the zodiac (hence its name). Look for it only when the sky is very dark, after evening twilight and before morning twilight. Like the planet Mercury, it appears near the sun, so the best times to see it are autumn evenings and spring mornings, although if you are at sea or in the mountains it can be seen during any season.

The zodiacal light is produced by sunlight scattered from tiny particles of cometary and meteoritic debris moving near the plane of the planets around the sun. These particles are gradually sinking toward the sun.

STAR CLUSTERS

Most stars were originally formed in large clusters containing hundreds, and occasionally thousands, of stars. The Pleiades ("Seven Sisters") is the brightest example of a star cluster in our sky, and it contains about five hundred stars, although only a handful are visible to the naked eye. Figure 10 shows the brighter members of this cluster, and the magnitudes of stars visible with binoculars are indicated. This is a very nearby example of an "open," or loose, cluster. Another example is the "double cluster" in the northern portion of the constellation Perseus, which may be picked out by naked eye on a dark night in fall; binoculars will clearly show that it is double, but they will not reveal its individual stars —they are too far. A third example is the Praesepe cluster in Cancer (M44).

If you scan the Milky Way with binoculars or a telescope, you will find dozens of faint, fuzzy patches, and many of these are open clusters too distant to show their individual stars.

Another type, the "globular" cluster, is tighter and more spherical in form. Unlike the open clusters, globular clusters avoid the plane of the Milky Way and are widely scattered through the galaxy, although concentrating toward its center. These differences are attributed to the fact that globular clusters were formed much earlier than open clusters, when the galaxy itself was more nearly spherical. The following examples are barely visible to the naked eye and may be picked out with binoculars: M13 in Hercules, M3 in Canes Venatici between the stars Cor Caroli and Arcturus, M22 in Sagittarius, and M4 in Scorpius. These clusters are indicated with an asterisk on the star finder.

THE ANDROMEDA GALAXY (M31)

This galaxy, very much like our own, lies at a distance of 2 million light-years outside our galaxy. It is the greatest system of stars visible to the naked eye—except our own.

To find the Andromeda galaxy, start from the northeast corner of the Great Square of Pegasus. Then go two stars farther to the northeast, and then step two stars to the north. Its location is indicated by a small square on the star finder. You will see a faint oval patch of light; this is the light of 100 billion stars!

Appendixes

Appendix One

Glossary of Astronomical Terms

Celestial Equator: The projection of the Earth's equator on the celestial sphere. This circle divides the sky into its northern and southern portions.

Celestial Pole: The point where the Earth's axis of daily rotation strikes the celestial sphere.

Celestial Sphere: An imaginary sphere against which we see the stars and on which we measure their positions.

Conjunction: Two objects are said to be *in conjunction* when they arrive at the same longitude on the ecliptic. *See* **Inferior Conjunction** and **Superior Conjunction.**

Constellation: A dot-to-dot picture in the sky. Most groups have no physical significance.

Declination: The celestial equivalent of latitude (*q.v.*). It is measured in degrees, and its symbol is delta, δ.

Earth: I have capitalized the name of our planet in this book because I am dealing with astronomy, not soil.

Ecliptic: The band of sky in which eclipses can occur, i.e., the path of the sun. It is also the path of the planets and is tilted 27° from the equator of the Earth.

Elongation: Angular distance from the sun. Mercury is visible only within a week or two of its greatest elongations.

Equator: The great circle halfway between the North and South Poles of the Earth.

Equinox: The instant when the sun crosses the celestial equator on its path around the ecliptic. Fall equinox occurs on September 21 and spring equinox on March 21, although the times vary slightly from year to year because our calendar has leap years.

Evening Sky: The half of the sky opposite the sun.

Focal Length: The distance between a lens and the image it makes of a distant object. You can determine the focal length of your camera lens by taking it out and casting an image onto a piece of paper—or by reading the label. (One inch is 25.4 millimeters.)

Galaxy: An "island universe," or an independent system of stars. Our Milky Way is one.

Inferior Conjunction: The instant when Mercury or Venus comes between us and the sun—more precisely, when one of them arrives at the sun's longitude and is on the near side of its orbit.

Latitude: A measure of distance from the equator, expressed as the angle between the equator and a point as seen from the center of the Earth. *See* **Declination.**

Longitude: A measure of distance from Greenwich, England, expressed as an angle along the equator from the meridian of Greenwich to the meridian of the point in question. *See* **Right Ascension.**

Magnification: The ratio of angular size with and without an optical device. With 10× magnification, an object appears ten times as broad. Equivalently, it appears to be at one-tenth the distance.

Magnitude: A measure of the faintness of a star. Sixth magnitude is the faintest we can ordinarily see, and the average of the twenty brightest stars defines the first magnitude.

Meridian: An observer's meridian passes through his zenith at the celestial poles. It is his north–south line.

Meteor: A flash of light produced by a particle in the atmosphere moving so rapidly that it becomes luminous.

Meteorite: A rock that once looked like a meteor.

Morning Sky: That half of the sky containing the sun.

Nebula: A cloud of gas or dust. A nebula may be luminous if excited by a nearby star, or it may appear dark against the background of starlight. In the past, this term was also used for galaxies because they appeared nebulous.

Opposition: The instant when one of the outer planets is directly opposite the sun. Because the orbits are fairly circular, this nearly coincides with the time of closest approach to the Earth.

Planet (literally, a "wanderer"): A solid object in motion about a star. If it is very small, it is called an asteroid.

Pole: The intersection of the Earth's axis of daily rotation with the surface of the Earth or the celestial sphere.

Precession of the Equinoxes: The Earth's axis of daily rotation slowly swings about the perpendicular to the sun's path. The motion is similar to that of a spinning top and is caused by the moon's gravitational pull on the Earth's equatorial bulge. The equinoxes slide around the ecliptic in 26,000 years.

Retrograde: Backwards or reversed. See the discussion of Mars's motion in Chapter 3. Jupiter and Saturn also show retrograde motions among the stars, but they are less marked.

Right Ascension: The equivalent of longitude on the celestial sphere. It is measured in hours from the spring equinox, where the ecliptic and equator intersect. The symbol for right ascension is alpha, α. One hour of right ascension is 15° of longitude.

Sky: The visible portion of the celestial sphere.

Spectrum: The color distribution of radiation from an object. Spectrum analysis permits a determination of chemical and physical composition.

Star Cluster: A group of stars within a galaxy. Some clusters contain dozens of stars; others contain millions.

Sundogs: Patches of light in the sky near the sun. They are produced by crystals of snow and ice high in the atmosphere.

Sunrise and **Sunset:** The moment when the sun's upper edge touches the imaginary horizon 90° from the zenith.

Superior Conjunction: The instant when a planet on the *far* side of the sun has the longitude of the sun. This is a time of invisibility, as is inferior conjunction.

Waning Crescent: Decreasing crescent. This occurs between full moon and new moon.

Waxing Crescent: Increasing crescent. This occurs when the moon is not waning.

Zenith: Straight up.

Zodiac: A band along the sun's annual path around the celestial sphere. (*See* **Ecliptic.**) The moon and planets remain in the zodiac.

The Greek Alphabet

LETTER	NAME	LETTER	NAME
α	Alpha	ν	Nu
β	Beta	ξ	Xi
γ	Gamma	o	Omicron
δ	Delta	π	Pi
ϵ	Epsilon	ρ	Rho
ζ	Zeta	σ	Sigma
η	Eta	τ	Tau
θ	Theta	υ	Upsilon
ι	Iota	ϕ	Phi
κ	Kappa	χ	Chi
λ	Lambda	ψ	Psi
μ	Mu	ω	Omega

Planetary Data

	ORBITAL PERIOD		MEAN DISTANCE FROM SUN	LEAST DISTANCE FROM EARTH	EQUATORIAL DIAMETER (MILES)	NUMBER OF KNOWN SATELLITES	MAXIMUM TEMPERATURE (FAHRENHEIT)
Mercury	88.0	days	.39	.55	3,010	0	780°
Venus	224.7	days	.72	.28	7,620	0	1000°
Earth	365.26	days	1.00	——	7,926	1	140°
Mars	687.0	days	1.52	.38	4,220	2	75°
Jupiter	11.9	years	5.20	4	88,800	12	−190°
Saturn	29.5	years	9.55	8	74,000	9	−270°
Uranus	84	years	19.20	17	29,500	5	−330°?
Neptune	165	years	30.09	29	27,200	2	−360°?
Pluto	248	years	39.5	29	3,600?	0	−360°?

Constellations on the Star Finder

NAME	ABBREVIATION	ENGLISH NAME	MYTHOLOGY
Andromeda	And	Andromeda	Chained woman; daughter of Cepheus and Cassiopeia.
Antlia	Ant	Air Pump	Modern.
Aquarius	Aqr	Water Carrier	
Aquila	Aql	Eagle	Armor-bearing bird of Jove.
Aries	Ari	Ram	Phrixus and his sister Helle fled his stepmother on the ram; Helle fell into the sea.

NAME	ABBREVIATION	ENGLISH NAME	MYTHOLOGY
Auriga	Aur	Charioteer	
Boötes	Boö	Bear Driver	
Cancer	Cnc	Crab	Banished to the sky for pinching the toe of Hercules.
Canis Major	CMa	Larger Dog	The dog of Orion.
Canis Minor	CMi	Smaller Dog	Orion's second dog.
Capricornus	Cap	Sea Goat	The goat-footed Pan; turned part fish upon plunging into the Nile to escape the monster Typhon.
Cassiopeia	Cas	Cassiopeia	Wife of Cepheus.
Cepheus	Cep	Cepheus	Father of the royal family.
Cetus	Cet	Whale	Sent to devour Andromeda, Cetus turned to stone at the sight of Medusa's head in the hand of Perseus.
Corona	CrB	Crown	Crown of Ariadne, daughter of King Minos.
Corvus	Crv	Crow	Sacred bird of Phoebus Apollo, who assumed bird's shape during battle.
Crater	Crt	Cup	The cup of Apollo.

NAME	ABBREVIATION	ENGLISH NAME	MYTHOLOGY
Cygnus	Cyg	Swan	Pet of Leda, the mother of Castor and Pollux, the Twins.
Delphinus	Del	Dolphin	The sacred fish that induced Amphitrite to become the wife of Neptune.
Draco	Dra	Dragon	The snake snatched by Minerva from the giants and flung to the sky.
Eridanus	Eri	River	Flowed into the Euxine Sea where Argonauts secured the golden fleece.
Gemini	Gem	Twins	Sons of Leda.
Hercules	Her	Hercules	Heroic laborer.
Hydra	Hya	Sea Serpent	Kept the crow away from Apollo's cup.
Leo	Leo	Lion	Slain by Hercules.
Libra	Lib	Scales	
Lupus	Lup	Wolf	
Lyra	Lyr	Lyre	Instrument of Orpheus, the musician of the Argonauts.
Monoceros	Mon	Unicorn	
Ophiuchus	Oph	Serpent Holder	Staff of Aesculapius, physician of the Argonauts.

NAME	ABBREVIATION	ENGLISH NAME	MYTHOLOGY
Orion	Ori	Orion	The hunter; placed in the sky opposite the Scorpion, by whose sting he was slain.
Pegasus	Peg	Pegasus	Winged horse; born from the blood of Medusa's head, severed by Perseus.
Perseus	Per	Perseus	Rescued Andromeda, using Medusa's head to turn the Whale to stone.
Pisces	Psc	Fish	
Puppis	Pup	Stern	Stern of the ship sailed by Jason and the Argonauts in search of the golden fleece.
Pyxis	Pyx	Compass	
Sagittarius	Sgr	Archer	
Scorpius	Sco	Scorpion	Slayer of Orion, who sets when the Scorpion rises.
Ursa Major	UMa	Great Bear	Juno, jealous of Callisto, transformed her to a bear.
Ursa Minor	UMi	Little Bear	
Virgo	Vir	Virgin	Proserpina, daughter of Ceres, abducted by Pluto.

The Thirty Brightest Stars

NAME		MAGNITUDE	POSITION		DATE OF 9 P.M. CULMINATION
1. α CMa	Sirius	−1.47	6h43m −16°39′		Feb. 15
2. α Car	Canopus	−0.73	6 21 −52 40		Feb. 10
3. α Lyr	Vega	+0.04	18 35 +38 44		Aug. 10
4. α Boö	Arcturus	0.06	14 13 +19 27		June 10
5. α Cen°		0.06	14 36 −60 38		June 15
6. β Ori	Rigel	0.08	5 12 − 8 15		Jan. 20
7. α Aur	Capella	0.09	5 13 +45 57		Jan. 20
8. α CMi	Procyon	0.34	7 37 + 5 21		Mar. 1
9. α Eri°	Achernar	0.47	1 36 −57 29		Dec. 1
10. β Cen°	Hadar	0.59	14 00 −60 08		June 5
11. α Aql	Altair	0.77	19 48 + 8 44		Sept. 1
12. α Ori	Betelgeuse	0.80	5 52 + 7 24		Feb. 1
13. α Tau	Aldebaran	0.86	4 33 +16 25		Jan. 15
14. α Vir	Spica	0.96	13 23 −10 54		May 25
15. α Cru°		1.05	12 24 −62 49		May 10
16. α Sco	Antares	1.08	16 26 −26 19		July 10
17. β Gem	Pollux	1.15	7 42 +28 09		Mar. 1
18. α PsA	Fomalhaut	1.16	22 55 −29 53		Oct. 15
19. β Cru°		1.24	12 45 −59 25		May 15
20. α Cyg	Deneb	1.26	20 40 +45 06		Sept. 15
21. α Leo	Regulus	1.35	10 06 +12 13		Apr. 9
22. ε CMa	Adhara	1.50	6 57 −28 54		Feb. 20
23. α Gem	Castor	1.58	7 31 +32 00		Mar. 1
24. γ Cru°		1.62	12 28 −56 50		May 10
25. λ Sco	Shaula	1.62	17 30 −37 04		July 25
26. γ Ori	Bellatrix	1.64	5 22 + 6 18		Jan. 25
27. β Tau	El Nath	1.65	5 23 +28 34		Jan. 25
28. β Car°	Miaplacidus	1.67	9 13 −69 31		Apr. 1
29. ε Ori	Alnilam	1.70	5 34 − 1 14		Jan. 28
30. ε Car°	Avior	1.85	8 21 −59 21		Mar. 10

	DISTANCE (LIGHT-YEARS)	TEMPERATURE (°K)	RADIUS (× SUN'S)	TRUE VISUAL BRIGHTNESS (× SUN'S)
1.	8.7	10,400	1.9	30
2.	100	7,500	35	5,700
3.	27	9,500	2.4	50
4.	38	4,250	35	95
5.	4.3	5,800	1.3	1.3
6.	500	11,200	80	60,000
7.	46	5,000	13	70
8.	11	6,450	2.6	12
9.	73	14,000	4.7	400
10.	190	22,000	22	10,000
11.	16	8,250	1.6	10
12.	300	3,800	1600	18,000
13.	64	3,900	90	110
14.	190	20,000	11	2,200
15.	220	18,000	4.9	3,000
16.	230	3,500	800	5,600
17.	33	4,100	19	73
18.	23	9,300	1.7	18
19.	500	26,600	43	4,400
20.	1600	9,500	100	63,000
21.	78	13,000	4	150
22.	330	21,000	16	5,000
23.	47	9,500	1.9	30
24.	60	3,000	200	800
25.	200	20,000	11	2,200
26.	230	21,000	11	2,500
27.	130	12,000	8	400
28.	80	9,500	3	65
29.	900	21,000	40	60,000
30.	330	4,100	100	600

* Invisible in North America and Europe.

Total and Annular Solar Eclipses

DATE		TYPE	VISIBILITY
1979	Feb. 26	Total	North America
	Aug. 22	Annular	Antarctica
1980	Feb. 16	Total	Central Africa, India, China
	Aug. 10	Annular	Pacific Ocean, South America
1981	Feb. 4	Annular	South Pacific
	July 31	Total	Black Sea, U.S.S.R., Japan

Lunar Eclipses

DATE			MID-ECLIPSE (E.S.T.)	TYPE
1978	Mar.	24	11:25 a.m.	Total
	Sept.	16	2:03 p.m.	Total
1979	Mar.	13	4:10 p.m.	Partial
	Sept.	6	5:54 a.m.	Total
1981	July	16	11:48 p.m.	Partial
1982	Jan.	9	2:56 p.m.	Total
	July	6	2:30 a.m.	Total
	Dec.	30	6:26 a.m.	Total
1983	June	25	3:25 a.m.	Partial

Events in the Sky

1978

Eclipses of the Moon
    Mar. 24, Sept. 16.

Twilight Configurations
    May 9–11, evening: Venus, Mars, Jupiter, and Saturn with new moon.
    June 7–9, evening: Repetition of May 9–11.
    July 7–9, evening: Mercury, Venus, Mars, and Saturn with new moon.
    Sept. 27–29, morning: Saturn and Jupiter with old moon.
    Dec. 25–27, morning: Mercury and Venus with old moon.

Conjunctions of Bright Planets
    June 4: Mars and Saturn.
    July 29: Venus and Saturn.

Aug. 13: Mars and Venus.
Oct. 23: Mars and Venus.
Meteor Showers with Favorable Moon
Jan. 3: Quadrantids.
May 5: Eta Aquarids.
July 28: Delta Aquarids.

1979

Eclipses of the Sun
Feb. 26, Aug. 22.
Eclipses of the Moon
Mar. 13, Sept. 6.
Twilight Configurations
April 23 and 24, morning: Mercury, Venus, and Mars with old moon.
July 26 and 27, evening: Jupiter and Saturn with new moon.
Sept. 16–19, morning: Mars, Jupiter, and Saturn with old moon.
Oct. 16–19, morning: Repetition of Sept. 16–19.
Nov. 15–17, morning: Repetition of Sept. 16–19.
Conjunctions of Bright Planets
May 20: Mars and Venus.
Dec. 15: Mars and Jupiter.
Meteor Showers with Favorable Moon
Jan. 3: Quadrantids.
July 28: Delta Aquarids.
Oct. 21: Orionids.
Dec. 12: Geminids.

1980

Eclipses of the Sun
Feb. 16, Aug. 10.
Twilight Configurations
May 16–18, evening; Venus, Mars, Jupiter, and Saturn with new moon.
June 14–16, evening: Mercury, Mars, Jupiter, and Saturn with new moon.

July 14–16, evening: Mars, Jupiter, and Saturn with new moon.

Aug. 6–8, morning: Mercury and Venus with old moon.

Aug. 12 and 13, evening: Mars, Jupiter, and Saturn with new moon.

Oct. 5–7, morning: Venus, Jupiter, and Saturn with old moon.

Oct. 10–13, evening: Mercury and Mars with new moon.

Conjunctions of Bright Planets

Feb. 26: Mars and Jupiter.

May 5: Mars and Jupiter.

June 23: Mars and Saturn.

Oct. 31: Venus and Jupiter.

Nov. 4: Venus and Saturn.

Dec. 31: Jupiter and Saturn.

Meteor Showers with Favorable Moon

Aug. 12: Perseids.

Dec. 12: Geminids.

1981

Eclipses of the Sun

Feb. 4, July 31.

Eclipse of the Moon

July 16.

Twilight Configurations

Aug. 2–4, evening: Venus, Jupiter, and Saturn with new moon.

Aug. 27 and 28, evening: Venus, Jupiter, and Saturn very close together.

Aug. 31 and Sept. 1, evening: Repetition of Aug. 2–4.

Sept. 18–22, evening: Mercury, Venus, Jupiter, and Saturn aligned in sky.

Nov. 2–5, morning: Mercury, Jupiter, and Saturn close together.

Nov. 23 and 24, morning: Mars, Jupiter, and Saturn with old moon.

Dec. 22–24, morning: Repetition of Nov. 23.

Conjunctions of Bright Planets

Mar. 4: Jupiter and Saturn.

Aug. 1: Jupiter and Saturn.

Aug. 27: Venus and Saturn.

Aug. 28: Venus and Jupiter.

Meteor Showers with Favorable Moon
      Jan. 3: Quadrantids.
      May 5: Eta Aquarids.
      July 28: Delta Aquarids.
      Oct. 21: Orionids.

Almanac of the
Planets

MERCURY

This planet is readily visible only within two weeks of the times of its greatest distance (elongation) from the sun, listed below.

The more favorable elongations are marked with an asterisk (*).

DATES OF GREATEST DISTANCES FROM THE SUN
(DISTANCES IN DEGREES)

	EAST (EVENING SKY)	WEST (MORNING SKY)
1978	Mar. 24 (19°)*	Jan. 11 (23°)
	July 21 (27°)	May 9 (26°)
	Nov. 16 (23°)	Sept. 4 (18°)*
		Dec. 24 (22°)

DATES OF GREATEST DISTANCES FROM THE SUN
(DISTANCES IN DEGREES)

	EAST (EVENING SKY)	WEST (MORNING SKY)
1979	Mar. 7 (18º)*	Apr. 21 (27º)
	July 3 (26º)	Aug. 19 (18º)*
	Oct. 29 (24º)	Dec. 7 (21º)
1980	Feb. 19 (18º)*	Apr. 2 (28º)
	June 14 (24º)	July 31 (19º)
	Oct. 10 (25º)	Nov. 19 (20º)*
1981	Feb. 1 (18º)*	Mar. 15 (28º)
	May 26 (23º)	July 14 (21º)
	Sept. 22 (26º)	Nov. 2 (19º)*

VENUS

Only the sun and moon are brighter than Venus, so a detailed description of its location among the stars is usually not needed. Clues to seeing Venus in the daytime are given in Chapter 3.

1978   *Invisible at the start of the year, Venus will be seen in the evening sky until November and in the morning from late November through the end of the year.*
Jan.   22: Passes behind the sun.
Mar.   5: Reappears in evening twilight near Mercury.
Aug.   29: Greatest eastward distance from the sun. Sets three hours after sunset.
Nov.   1: Vanishes in evening twilight near Mercury.
Nov.   15: Reappears in morning twilight.

1979   *Visible in the morning sky through the spring. Invisible during summer and early autumn.*

Jan. 18: Greatest westward distance from the sun. Rises three hours before sunrise.
July 15: Vanishes into morning twilight.
Aug. 24: Behind the sun.
Oct.   3: Reappears in evening twilight below Mercury.

1980   *Visible in the evening sky through spring and the morning sky during summer and fall.*

June 10: Vanishes into evening twilight.
June 15: Passes in front of the sun.
June 24: Reappears in morning twilight.
Aug. 25: Greatest westward distance from the sun. Rises three hours before sunrise.

1981   *Visible in the morning sky during winter and the evening sky from late spring through the end of the year.*

Feb. 24: Vanishes into morning twilight.
Apr. 10: Passes behind the sun.
May 20: Reappears in evening twilight near Mercury.
Nov. 10: Greatest eastward distance from the sun. Sets three hours after sunset.

MARS

The red planet will be prominent in the evening sky during 1978 and 1980. A series of three conjunctions with Jupiter occurs in 1979 and 1980.

1978   *Mars begins the year rising at sunset, and it will be visible in the evening sky until late November, when it vanishes into twilight. In Cancer until May, when it moves into Leo and then eastward along the zodiac to Scorpio in November.*

Jan. 21: Opposite the sun.
Mar.   2: Ends westward (retrograde) motion among the stars.
Apr. 26: On the meridian at sunset.
June   4: Conjunction with Saturn in the evening sky.
Dec.   1: Vanishes into evening twilight.

1979 *Mars will be in front of the sun and invisible until mid-March, when it reappears in morning twilight. Rises earlier each month and visible in the morning sky through the end of the year. In Pisces until mid-May, then moving eastward to Leo in December for the first of an unusual series of three conjunctions with Jupiter.*

Jan. 16: Behind the sun.
Mar. 17: Reappears in morning twilight below Venus.
May 20: Conjunction with Venus.
Nov. 23: Rises at midnight.
Dec. 15: The first of three conjunctions with Jupiter.

1980 *Starting the year in the morning sky, Mars moves to the evening sky in February and remains visible through the year. In Leo until June 20, then eastward to Sagittarius at the end of November.*

Jan. 15: Begins westward motion among the stars.
Feb. 24: Opposite the sun, rising at sunset.
Feb. 26: Second conjunction with Jupiter.
Apr. 10: Ends westward motion.
May 5: Third conjunction with Jupiter.
May 30: On the meridian at sunset.
June 23: Conjunction with Saturn.

1981 *Mars will be seen briefly in evening twilight early in January, be lost in twilight, and reappear in the morning sky in May. Mars moves from Taurus in May eastward to Virgo in December.*

Apr. 3: Behind the sun.
July 4: Near Mercury in morning twilight.
Dec. 23: Rises at midnight.

JUPITER

During the next few years, this planet will be favorably located from late fall to early spring as it approaches Saturn. These two planets go through an extremely rare series of three conjunctions in 1980 and 1981. Jupiter also has a series of conjunctions with Mars in 1979–1980.

1978 *Rises near sunset in January and will be seen in the evening sky, setting earlier each month, until mid-June, when it vanishes into twilight. From August through the end of the year, it may be seen in the morning sky. In Gemini until August, and then in Cancer.*

Mar. 10: On the meridian at sunset.

July 10: Passes behind the sun.

Nov. 2: On the meridian at sunrise.

1979 *Rises near sunset in January and February, and then visible in the evening sky until it vanishes into twilight late in July. From September through the end of the year, it will be visible in the morning sky. In Cancer until August, and then in Leo.*

Jan. 24: Opposite the sun.

Apr. 20: On the meridian at sunset.

Aug. 13: Behind the sun.

Dec. 2: On the meridian at sunrise.

Dec. 15: Conjunction with Mars.

1980 *Well located near Mars and Saturn in the evening sky through mid-August, when it vanishes into evening twilight. Visible in the morning sky from October through the end of the year. In Leo until September, then Virgo. Rare series of conjunctions with Saturn begins at the end of the year.*

Feb. 24: Opposite the sun.

Feb. 26: Conjunction with Mars.

May 5: Conjunction with Mars.

May 21: On the meridian at sunset.

Sept. 12: Behind the sun.

Dec. 31: Conjunction with Saturn. On the meridian at sunset.

1981 *Jupiter remains in the evening sky very near Saturn until late September, when it vanishes into twilight. During November and December, it will be visible in the morning sky east of Saturn. In Virgo all year.*

Mar. 4: Conjunction with Saturn.

Mar. 25: Opposite the sun, rising at sunset.

June 23: On the meridian at sunset.

Aug. 1: Final conjunction with Saturn.

Oct. 13: Behind the sun.

Nov. 1: Reappears in morning twilight.

SATURN

Well placed in winter and spring for the next few years. Saturn's rings will appear relatively narrow and they may vanish briefly in late 1979 and early 1980 when they line up with the Earth. See JUPITER concerning a very rare series of conjunctions.

1978   *Visible in the evening until it vanishes into twilight early in August. Reappears in mid-September in the morning sky. In Leo all year.*
        Feb.   10: Opposite the sun, rising at sunset.
        May   14: On the meridian at sunset.
        June   4: Conjunction with Mars in evening sky.
        Aug. 28: Behind the sun.
        Dec.   6: On the meridian at sunrise.

1979   *Visible in the evening until it vanishes into twilight late in August. Reappears in the morning sky near the end of September. Rings may vanish temporarily in mid-October. In Leo east of Jupiter all year.*
        Mar.   2: Opposite the sun, rising at sunset.
        May   28: On the meridian at sunset.
        Sept. 11: Behind the sun.
        Dec. 19: On the meridian at sunrise.

1980   *Rising after midnight in January, Saturn moves westward into the evening sky in March, where it will be seen near Mars and Jupiter until vanishing into twilight early in September. Reappears in morning twilight east of Jupiter early in October. Rings may be very difficult to see from April through June, when they again line up with the Earth. In Leo until July, then Virgo.*
        Mar. 15: Opposite the sun.
        June 11: On the meridian at sunset.
        June 23: Conjunction with Mars.
        Sept. 22: Passes behind the sun into the morning sky.
        Dec. 31: First conjunction with Jupiter.

1981   *Saturn remains unusually close to Jupiter throughout the year as the planets go through a rare series of conjunctions. For the location of Saturn, see JUPITER for this year.*

Phases of the Moon

	1978				1979		
	NEW		FULL		NEW		FULL
Jan.	9^d	04^h*	24^d	08^h	Jan.	28^d 06^h	13^d 07^h
Feb.	7	15	23	01	Feb.	26 16	12 01
Mar.	9	02	24	15	Mar.	28 02	13 21
Apr.	7	14	23	03	Apr.	26 12	12 12
May	7	04	22	12	May	25 23	12 01
June	5	18	20	20	June	24 12	10 11
July	5	09	20	03	July	24 01	9 20
Aug.	4	01	18	11	Aug.	22 17	8 03
Sept.	2	16	16	19	Sept.	21 10	6 11
Oct.	2	07	16	06	Oct.	21 02	5 20
Oct.	31	20	—	—	Nov.	19 18	4 06
Nov.	30	08	14	20	Dec.	19 08	3 18
Dec.	29	19	14	12			

* The times given are days and hours at the longitude of Greenwich, England. For U.S. time zones, subtract 5 hours for E.S.T., 6 hours for C.S.T., etc., or 4 hours for E.D.T., 5 hours for C.D.T., etc.

	1980				1981			
	NEW		**FULL**			**NEW**		**FULL**	
Jan.	17^d	21^h	2^d	09^h	Jan.	6^d	07^h	20^d	07^h
Feb.	16	08	1	02	Feb.	4	21	18	22
Mar.	16	18	1	20	Mar.	6	09	20	14
Mar.	—	—	31	12	Apr.	4	19	19	07
Apr.	15	03	30	07	May	4	03	18	22
May	14	11	29	21	June	2	11	17	15
June	12	21	28	09	July	1	19	17	04
July	12	07	27	18	July	31	04	—	—
Aug.	10	19	26	03	Aug.	29	14	15	16
Sept.	9	09	24	12	Sept.	28	04	14	03
Oct.	9	02	23	21	Oct.	27	20	13	13
Nov.	7	21	22	07	Nov.	26	15	11	23
Dec.	7	15	21	18	Dec.	26	10	11	09

Using the Star Finder – Locater Wheel

The star finder will permit you to visualize the positions of the stars above your horizon at any hour of any night or day. To use the star finder, rotate the disk until your date appears next to the hour (Standard Time). Hold the star finder over your head and turn it so the point marked "south" is toward your south. You will then see the stars in their positions at the time within the "horizon."

The star finder also has other uses. For example, you may determine the time of sunset (or sunrise). To do this, note that the sun's path among the stars is indicated by a band with dates alongside. Each date indicates the corresponding position of the sun. Turn the disk until the sun's position for your date falls on the western horizon (eastern horizon for sunrise). Now, on the outer edge of the star finder read the hour corresponding to your date. This is the hour of sunset. When the star finder is set this way, it will also help you find the bright stars when they appear in the twilight.

See the index for references to other uses of the star finder.

Index

A NOTE ABOUT THE AUTHOR

Charles Allen Whitney was born in Milwaukee, Wisconsin, in 1929. After receiving a B.S. in physics from the Massachusetts Institute of Technology, Whitney completed his M.A. and Ph.D. in astronomy at Harvard University. He has been a physicist at the Smithsonian Astrophysical Observatory since 1956, and is also a professor of astronomy at Harvard. The Discovery of Our Galaxy *was published by Knopf in 1971. He is married, the father of five children, and lives in Weston, Massachusetts.*

A NOTE ON THE TYPE

This book was set in Caledonia, a Linotype face designed by W. A. Dwiggins. It belongs to the family of printing types called "modern face" by printers—a term used to mark the change in style of type letters that occurred about 1800. Caledonia borders on the general design of Scotch Modern, but is more freely drawn than that letter.

The book was composed, printed, and bound by The Book Press, Brattleboro, Vermont. Star finder manufactured by Creative Lithographers, New York, New York. Typography by The Etheredges.